Palgrave Studies in Literature, Science and Medicine

Palgrave Studies in Literature, Science and Medicine is an exciting new series that focuses on one of the most vibrant and interdisciplinary areas in literary studies: the intersection of literature, science and medicine. Comprised of academic monographs, essay collections, and Palgrave Pivot books, the series will emphasize a historical approach to its subjects, in conjunction with a range of other theoretical approaches. The series will cover all aspects of this rich and varied field and is open to new and emerging topics as well as established ones.

More information about this series at
http://www.palgrave.com/gp/series/14613

Anna Gasperini

Nineteenth Century Popular Fiction, Medicine and Anatomy

The Victorian Penny Blood and the 1832 Anatomy Act

palgrave
macmillan

Anna Gasperini
Independent Scholar
Perugia, Italy

Palgrave Studies in Literature, Science and Medicine
ISBN 978-3-030-10915-8 ISBN 978-3-030-10916-5 (eBook)
https://doi.org/10.1007/978-3-030-10916-5

Library of Congress Control Number: 2018967281

This Palgrave Macmillan imprint is published by the registered company Springer Nature Switzerland AG
The registered company address is: Gewerbestrasse 11, 6330 Cham, Switzerland

To my husband. Thank you.

[A]natomy is the very basis of surgery [...] [It] informs the head, *gives dexterity to the* hand, *and familiarizes the* heart *with a sort of necessary inhumanity [...].*
William Hunter, anatomist, 1764

Well, for my part [...] I think it's wery hard if, after paying rates and taxes [...] I should be obleeged to go to the workhus, and then be cut up in a surgeon's slaughterhouse at last.
Poor widow Mrs. Smith, *The Mysteries of London*, 1845

Horror is about trying to codify anxiety, trying to name and understand those things we fear.
John Logan, creator of horror TV show *Penny Dreadful*, 2014

Preface

Dissecting a Literary Monster: Why?

We are used to thinking about monsters as frightening: grotesque assemblages of malformed parts, often huge, possibly supernatural, and certainly malevolent. Yet, if one looks up the word 'monster' in the dictionary, he/she will discover a more nuanced meaning which consists, simultaneously, of 'frightening', 'huge', and 'marvellous'.[1] In this multifaceted sense, the penny blood genre is a literary monster: a gargantuan combination of scattered pieces from those cultural forms that did not have a place in mainstream knowledge, an abomination for those social strata that could not control it, but wonderful for the masses that in the 1830s and 1840s were discovering the pleasure of leisure reading.

For a long time, the penny bloods and penny dreadfuls were all but forgotten, a mythical creature barely mentioned as something comparable to the monstrous hybrid of gothic novels and better forms of serialized popular fiction. This rather unkind perception stemmed from the original Victorian middle-class viewpoint on the penny bloods, which perceived this subversive literary form—violent, licentious, almost freely available to the working-class and, most of all, beyond their control—as dangerous. This negative narrative had long-lasting consequences. Until relatively recently, it influenced academic judgement of the penny bloods' importance as a literary form, crucially impacting on the production and circulation of knowledge about them. It could be said that, for the best part of their posthumous life so far, penny bloods have been considered the eccentric relatives of the Victorian literature family: seldom included in formal gathering invitations, no one would willingly acknowledge any closeness to them, nor discuss their outlandish quirks in too much detail, if at all. The prejudice started decreasing thanks to the early efforts of such scholars as Edward S. Turner and Louis James, who first analysed the scarce original material available combining skilful book history research with study of what little information had trickled through the merciless sieve of nineteenth-century cultural commentators. Later, scholars such as Anne Humpherys, John Springhall, Helen R. Smith, and Robert L. Mack, among others, did impressive work cataloguing collections and analysing authors and narratives, contributing to the gradual rediscovery of this fascinating, but still comparatively underexplored, corner of Victorian fiction. This book contributes to this operation of rediscovery. I started working on penny bloods and penny dreadfuls almost by chance, having never worked on serialized fiction before. After a first puzzled moment in which I realized that they were different from any other literary form I had encountered so far (meaning they were much longer, they rambled, and they made absolutely no sense if one insisted on reading them as novels), I started formulating the idea that their relationship with their world—their readership's world—was a complex one that involved facts and people belonging to a variety of spheres, some of them rather unexpected.

On Monday 28 April 1828, the Select Committee on Anatomy, appointed by the House of Commons to investigate the matter of how anatomy schools obtained bodies for dissection, started its hearings. The very first witness had an eminent name: it was Sir Astley Cooper, Bart., president of the Royal College of Surgeons, who had acquired his title after successfully removing a cyst from the sovereign's scalp. He was also one of the most prominent anatomists in the kingdom. The Committee's proceedings, collected in the *Report from the Select Committee on Anatomy*, abound in important names from the world of medicine in the first decades of the century: John Abernethy, Thomas Southwood Smith, and Thomas Wakley, among others. On Friday 2 May, though, the last but one hearing is simply marked as that of 'A.B.'. To this day, we do not know this man's identity, although there are speculations.[2] His name could not be recorded, partly to protect him, but also—and most importantly—to protect the medical gentlemen whose names appear in the *Report* from association with him. Compared to the elaborate wording of the men who preceded A.B., his answers to the Committee are brief, laconic, and dry. The people in the room despised him, but this man was essential to dissection activities in the city of London.

'Is it not', the Commission asked him, 'your occupation to obtain bodies for anatomical schools?' 'Yes,' A.B. replied, 'it has been for some years'.

A.B. was a bodysnatcher.

People in his line of business, for this is what it was, stole bodies from fresh graves at night, and/or obtained fresh bodies by other illicit means[3] to sell it to anatomists for dissection. Mostly, they stole the bodies of the poor, whose tombs were an easier target than those of the better-off. It was a remunerative trade—too much so for the anatomists, who were paying increasingly dear prices for 'dissection material' sold by night in the antechambers of their dissection rooms, as the swelling numbers of medical students required more and more bodies. They had been asking for an alternative, legal supply of 'subjects' since the early years of the century and, finally, sympathetic political factions were listening to their pleas, hence the hearings of the Select Committee.

The first time I read the word 'bodysnatcher', though, was not in the *Report of the Select Committee for Anatomy*, but rather in a penny blood, *Varney the Vampyre; or: the Feast of Blood*. As a penny blood, *Varney* is a gory, lurid, aesthetically limited serialized story interspersed with supernatural events, hidden treasures, hanged bodies, and fearful dungeons. Interestingly, it also features experiments on cadavers: Varney's body is galvanized back to life by a medical student in an episode that is redolent of Mary Shelley's *Frankenstein*. Much later in the narrative, another body disappears from its grave, arising suspicions of bodysnatching. As I explored this genre, I found that resurrected bodies, bodysnatchers, and unscrupulous doctors appeared in various penny bloods in a sort of recurring *danse macabre*, the repetition of which did not only suggest interest in this triad of figures, but also a common origin in an event, in a heartfelt reality. The parallel historical research I conducted to better understand the nature and working of the bodysnatching business, in which the three elements appeared to be all simultaneously involved, led me to the 1832 Anatomy Act.

During his hearing, A.B. was asked: 'Suppose the bodies of those who die in workhouses, and have no friends to claim them, were given up, do you think that the public would be much against that practice?' To which A.B. replied that, after being initially 'prejudiced', people would finally 'come round' and accept it.[4] The ultimate goal of the Select Committee on Anatomy was, indeed, the passing of a law that allowed anatomists to source dissection material from the pauper. When it was passed, the Anatomy Act imposed that the bodies of those who were too poor, or whose families were too poor, to afford a funeral were to be handed over to anatomy schools for dissection. In the early phases of my research, I noted that the Act was passed in the same decade in which the serialized publications market was gathering momentum, and that the law impacted on the same constituency to whom the lurid penny blood series catered—the working class. It's been more than thirty years since the repercussions of the Anatomy Act on the Victorian pauper were first problematized[5]: this momentous change in medical history provided a first system to supply medical students with plenty of material to understand the functioning of human anatomy; however, it impacted on an underrepresented, powerless social category, dramatically changing their experience of death, which was already harrowing. A funeral, even one including the bare minimum paraphernalia, was no small expense for a poor family, which meant that a corpse was laid in the same environment where the family lived until they could raise the necessary sum. Besides, paupers' coffins—cheap, fragile, often heaped in piles in common pits and covered with a thin layer of ground—were already the most exposed to the resurrectioni sts' incursions. The Act, therefore, sourced bodies from a category that was already subjected to the violation of their post-death bodily integrity, unprotected by the law, and who perceived dissection as an ignominious fate, having it been part of murderers' death sentences since 1745. Furthermore, as A.B. was being interviewed by the Select Committee in London, another drama was being consumed in Edinburgh, which would explode as a scandal the following October and show the true extent to which the paupers were exposed by the increasing demands of the anatomy schools.

The Burke and Hare homicides in Edinburgh shook the world of anatomy and the public opinion. The murderers targeted vagrants, prostitutes, strangers in need of a shelter for the night—in a word: paupers. The channels through which anatomy schools obtained bodies, and the anatomists' carelessness as to where the material came from, were the object of unprecedented scrutiny. The horrible possibilities of 'burking', that is, murdering with a view to selling the body for dissection, generated a mass hysteria that was rekindled when the London burkers case, also known as the 'Italian Boy' case, exploded in the capital in 1831. Again, the known victims of Bishop and Williams, the murderers, were an Italian boy who roamed the metropolis showing a cage of white mice, an English homeless boy, and a single mother without a stable income. Paupers. The promoters of the Anatomy Act used this second case to speed up the passing of the law through Parliament. Mere months after Bishop and Williams had been hanged and dissected for murdering paupers and selling them for dissection, the Act sanctioned the use of the poorest among the poor as anatomical material.

Looking at these facts as immediately preceding the rise of penny blood fiction, its interest in resurrectionism, burking, stolen corpses stretched under the moonlight, and people disappearing in the urban underground assumes a different nuance. Of course, it could be argued that this interest is merely in tune with the bloods' sensationalistic, swashbuckling style. After all, young George Augustus Sala, working as an illustrator for Edward Lloyd, one of the chief publishers of penny blood series, once received a letter from his employer asking him to improve his work: 'The eyes [...] must be larger; and there must be blood – much more blood!'[6] The blood's famously repetitive plots may be easily advanced as a further reason for the recurrent macabre burking/resurrectionism stories. However, this is not enough explanation when we consider that to penny blood readers the subject was not merely a breath-taking story: for them dissection, bodysnatching, and forfeiture of one's body to the anatomists after 48 hours under the Anatomy Act were a terrifying reality. Granted, burkers and resurrectionists were part of the sensational lore of the country through the press' treatment of the Edinburgh and London murders and the occasional coverage of a bodysnatcher attack to a cemetery. On the other hand, from newspapers and trial reports, we know that they were also the object of a deep-rooted hatred on the part of the masses: burkers were despised and their relatives attacked, bodysnatchers and their accomplices were likely to incur in bodily harm if discovered, and hospitals, and medical students themselves, were not popular either. Historians have amply demonstrated that the passing of the Anatomy Act generated profound distress in the poor, which did not receive much attention in its own time, but can be reconstructed through careful rereading of records.[7] Penny bloods' ominous *danse macabre* must have had a different purpose than mere peddling of cheap sensation, and this book is about understanding what this purpose might have been.

This operation is necessary now more than ever, especially considering that, while the genre is gradually becoming popular among scholars, it is also attracting the attention of an entirely different field, that is, mainstream popular culture. Between 2014 and 2016, the horror TV show *Penny Dreadful* brought the genre under the spotlight. Series creator John Logan explained that the fascination Gothic narratives has been arising in people since they were written, stating that 'horror is about trying to codify anxiety, trying to name and understand those things we fear'.[8] What Logan suggests is that horror and gore in a narrative are never an end in themselves, but they express the deep-seated fears of an age, of an audience. It is my firm belief, after almost eight years spent studying penny bloods that this applies to them too, to the original serialized horror narratives circulated through the mass publication industry. Consequently, times are mature for further exploring what fears the violence and terrors that characterized penny blood narratives were trying to elaborate.

This book goes deeper in dissecting the literary monster, considering a group of highly successful penny bloods to map out what anxieties they wanted to 'name and understand', in Logan's words, for their readership. Now that news of this almost-forgotten genre is reaching the wider public, and nowadays popular fiction is reusing it, it is crucial that we achieve a clearer understanding of how penny bloods dialogued with their contemporary reality, with their readers' everyday life, which was made also of political, scientific, and social changes. This genre's interaction with their readership's context was more sophisticated than it is still, to an extent, credited for and included a range of meanings spanning from discourse circulation, to ethics, to bodily agency issues. Gaining a stronger grasp of this interaction will increase our appreciation, not only of the nuances of lower class culture, but also of early forms of literary elaboration of anxieties about scientific developments. There seems to be a gap in literary history between *Frankenstein*'s bodysnatching doctor and his tragic monster and the disquieting clinicians of sensation fiction by Wilkie Collins and others. In fact, there was no gap: anxiety about medicine's invasion of one's own bodily boundaries was indeed discussed, only not in mainstream literature. It was the literature of the pauper, the constituency that was more likely to experience the touch of the lancet on the skin and yet was unable to voice its fears that originally discussed anxieties about the burgeoning field of anatomy. This book therefore, 'dissects' a literary monster in the sense that it uncovers a hidden complex system of narratives, themes, and characters, whose frame is traceable in subsequent forms of nineteenth-century popular fiction, such as sensation fiction, as well as in nowadays forms of popular fiction. Understanding this frame will lead, ultimately, to greater awareness of its survival, of its evolution into our own uses of the serialized horror formula to 'codify' our own anxieties.

Perugia, Italy Anna Gasperini

Notes

1. ('monster', Def. 6, 2, 4b), *Oxford English Dictionary*, n.d.
2. Sarah Wise suggests that it could be John Bishop, occasional resurrectionist and 'burker', tried and hanged for the murder of three people to sell them as dissection subjects in 1831. See Sarah Wise, *The Italian Boy—Murder and Grave-Robbing in 1830s London*, Pimlico 20 (London: Jonathan Cape, 2004), 36.
3. Such as pretending to be a relative of a deceased poor in a workhouse, or simply stealing the corpse from the house in which it was being waked.
4. 'Report from the Select Committee for Anatomy' (London, 1828), 72.
5. In 1987, Ruth Richardson published her groundbreaking study *Death, Dissection and the Destitute*, still the pillar of studies focusing on the impact of the Anatomy Act on the Victorian poor.
6. George Augustus Sala, *The Life and Adventures of George Augustus Sala, Vol. 1* (New York: C. Scribner's Sons, 1895), 209.
7. See, for instance, Ruth Richardson's analysis of workhouse inmates' cannibalistic anxieties in the 'nattomy soup' incident in Ruth Richardson, *Death, Dissection and the Destitute* (Chicago and London: The University of Chicago Press, 1987), 221–22. Another example is Elizabeth T. Hurren's examination of the Mary Ann Huckle case, in which Huckle appeared in court for withholding her husband's body for fear of losing his body to the anatomists, in Elizabeth T. Hurren, *Dying for Victorian Medicine—English Anatomy and Its Trade in the Dead Poor, c. 1834–1929* (London: Palgrave Macmillan, 2012), 205–6. Both episodes are examined in the next chapters.
8. Showtime, '#2: Literary Roots,' *Penny Dreadful Production Blog*, 2014, http://www.sho.com/video/24671/dreadful-production-blog-2--literary-roots.

List of Works Cited

'('monster', Def. 6, 2, 4b).' *Oxford English Dictionary*, n.d.

Hurren, Elizabeth T. *Dying for Victorian Medicine—English Anatomy and Its Trade in the Dead Poor, c. 1834–1929*. London: Palgrave Macmillan, 2012.

'Report from the Select Committee for Anatomy.' London, 1828.
Richardson, Ruth. *Death, Dissection and the Destitute*. Chicago and London: The University of Chicago Press, 1987.
Sala, George Augustus. *The Life and Adventures of George Augustus Sala, Vol. 1*. New York: C. Scribner's Sons, 1895.
Showtime. '#2: Literary Roots.' *Penny Dreadful Production Blog*, 2014. http://www.sho.com/video/24671/dreadful-production-blog-2--literary-roots.
Wise, Sarah. *The Italian Boy—Murder and Grave-Robbing in 1830s London*. Pimlico 20. London: Jonathan Cape, 2004.

Acknowledgements

Books take research. Research takes effort. Efforts take people who make the effort bearable and help you growing a project out of an idea. I have quite a few to thank for their support with this particular effort, and here they are.

First and foremost, Dr. Elizabeth Tilley from NUI Galway, my Ph.D. supervisor. She believed in me and my project from day one. I could not have hoped for best direction when I first approached academic research. Still from NUIG, Professor Richard Pearson, for guiding me through the wonderful world of the Victorian literary marketplace and for being a great advisor during my years in Galway. Finally, Dr. Muireann O'Cinneide from NUIG and Professor Darryl Jones from Trinity College Dublin, my Viva Voce examiners, for first encouraging me to create and publish this book.

I am indebted to numerous libraries and archives that have kindly allowed me to use their material: the National Library of Scotland, the British Museum, the British Library, Yale University Historical Library—Cushing/Whitney Medical Library, and Glasgow University Library, for agreeing to let me use images from their archives, and the Wellcome Library, for making theirs available through the Creative Commons Attribution (CC-BY) license.

There are several extraordinary academics who helped me through my research process. I would like to thank Professor Emeritus Louis James, Dr. Elizabeth T. Hurren, Professor John Spiers, and Dr. Robert L. Mack: with their work, and by taking time to talk and give advice to a Ph.D. student who was navigating a complex world, they helped me to complete this research. My gratitude goes also to Professor Rohan McWilliam, who encourages my research and my work, and who offered kind and constructive advice during the stages of my publication proposal submission. I would also like to thank the academic community at Victoria Listserv and the Victorian Popular Fiction Association Organisers, Dr. Janine Hatter, Dr. Helena Ifill, and Dr. Jane Jordan. Last, but definitely not least, a special mention goes to Dr. Ruth Richardson and Ms. Sarah Wise: with your work, you inspired me and instructed me; with your kind words, you fostered my passion. Without you, this book would not be here.

I would also like to thank the staff at Palgrave Macmillan for making this possible: Ben Doyle, publisher, for believing in this book from the first email, and Camille Davies, indefatigable editorial assistant, for her help and professionalism.

I cannot possibly forget the people who support me every day with their love: to Andrea, my wonderful husband, for whom there are not words enough; to mum and dad, for first taking me to London and basically for everything else; to my brother and sister, who are there, always. I am a very lucky person indeed.

From the bottom of my heart, thank you.

Perugia, Italy
September 2018

Contents

List of Figures

CHAPTER 1

The Subject Examined: Penny Bloods, the Anatomy Act, and a Common Ground for Analysis

1 New Readers and Alternative Culture: Rise of a Literary Monster

In 1874, social journalist James Greenwood published the collection of essays *The Wilds of London*, containing a series of pieces authored by him on lower class life in the burgeoning metropolis. About halfway through it, there is the essay 'A Short Way to Newgate'. The ominous title suggests that he will discuss some disgraceful practice that culminates in reclusion in the famous prison. From his point of view, it was so. 'There is a plague' he solemnly announces 'that is striking its upas roots deeper and deeper into the English soil [...] bringing death and misery unspeakable'.[1] As the page progresses, and Greenwood paints an increasingly murky portrait of this 'plague', the reader may think that he is actually about to discuss a new and terrible disease propagated by some evil plague-spreader, until he dramatically announces that this plague is 'poisonous literature'.[2]

He is talking about penny dreadfuls.[3]

With what reads almost like personal grudge, he fulminates for fourteen pages against them, declaring that the 'virulent' penny dreadfuls, made all the more dangerous by their cheap price that facilitates their 'propagation',[4] are 'poison[ing] the minds of boys and girls'.[5] He even aims a three-page vitriolic attack to G.W.M. Reynolds himself, one of the fathers of the genre, accusing him of corrupting innocent young minds.[6]

A. Gasperini, *Nineteenth Century Popular Fiction, Medicine and Anatomy*, Palgrave Studies in Literature, Science and Medicine,
https://doi.org/10.1007/978-3-030-10916-5_1

Greenwood's perspective was biased, and understandably so: although, as a social reformer, he was interested in working-class life, his perception of penny dreadfuls was the quintessentially middle-class one of a man bent upon 'improving' the lower class, on guiding them towards what he believed was healthy for them. Working-class readers, by contrast, resisted all middle-class attempts to 'protect' them from the penny bloods and penny dreadfuls. This literature was necessarily different: from the viewpoint of mainstream literature, which was for the exclusive use of the upper classes (who could afford books), its style and aesthetic were disgraceful; worst of all, it was tailored on working-class culture, which conceived a more violent and less restrained idea of entertainment.[7] Indeed, the best way to understand the penny blood as a literary object is by looking at it as the product of working-class alternative culture combined with a set of historical coincidences.

Although unwittingly, the middle class played a part in the rise of this literary monster. After the 1790s Jacobin panic, they thought that it wise to provide the masses with just enough education as to make them more manageable.[8] Initiating them to leisure reading was never the plan; yet, the introduction of basic literacy skills yielded an unexpected, but inevitable, result: the working class developed a taste for reading. Simultaneously, the book market, which the Napoleonic wars had severely impaired, was finding new life through serialized publications. 'Serial', as Graham Law and Robert L. Patten observe, is indeed a neologism from the 1830s that marks the momentous changes the invention of the rotary steam press and the paper-making machine triggered in the publication industry.[9] It is 1840; the population has been steadily increasing since the beginning of the century, and so have literacy rates.[10] From Salisbury Square, just off Fleet Street, down by the northern bank of the Thames, Edward Lloyd, publisher of cheap serialized literature and nemesis of 'respectable' authors and publishers, answers the demand created by the huge pool of new readers hungry for fiction: the first instalment of Thomas Peckett Prest's penny blood series *Ela, the Outcast, or: the Gipsy of Rosemary Dell* sees the light.

Initially, the serialized publication market was essentially middle class in nature. In the 1830s, Dickens, soon followed by other popular authors, made respectable the originally lower class monthly parts form, publishing his novels in monthly issues. Although cheaper than complete books, 1*s*. was still too high a price for working-class readers.[11] A first form of serialized publication addressing them came from the religious tracts societies, who saw in this rising market a new tool to aid them in their mission: improving the masses. The religious tracts, mass-produced and distributed home to home for free by the volunteers of various evangelical societies, contained edifying pieces devoted to the enlightenment of the poor. While the role of religious tracts in promoting literacy among the working class must not be underestimated,[12] the truth is that they were inadequate to meet their target audience's tastes. Both the tracts and their distributors conveyed a patronizing message dramatically unsuitable to the political climate, they reinforced the traditional social hierarchy, and were often insensitive towards the reality of the poor readers.[13] Evangelicals would offer the religious tracts to the poor as a relief against hunger or in view of cholera outbreaks, and preached frugality and diligence to a starving, jobless class.[14] Dickens famously ridiculed this attitude in *Bleak House* with Mrs. Pardiggle, who behaves as 'an inexorable moral Policeman' and distributes reading material so appalling that Mr. Jarndyce 'doubted if Robinson Crusoe could have read it, though he had no other in his desolate island'.[15] This last remark pinpoints one of the tracts' greatest shortcomings, from the intended readers' point of view, namely: they did not contain fiction. Evangelicals were deeply suspicious of fiction and preferred to distribute 'didactic' material or biographies of suitably 'pious individuals'.[16] A similar love for truth and facts, and therefore lack of fiction, characterized the literature with which the secular S.D.U.K. provided the libraries attached to the mechanics' institutes, which, though plentiful, was strictly devoted to 'useful knowledge'.[17] In brief, societies bent upon the improvement of the masses refused to acknowledge the demand for fiction from their audience, whose tastes were 'diverse and encompassing, [...] never exclusive, and embraced several cultural levels simultaneously'.[18]

The premises for developing fiction that specifically appealed to this public were already in place: Simon Eliot notes that the concept of cheap serialized publishing for the poor already existed in the eighteenth-century tradition of chapbooks, and in the early-nineteenth-century forms of the broadside ballad and the boxiana.[19] Furthermore, the late-eighteenth- and early-nineteenth-century Gothic mania, with its last outrageous offspring, the Minerva Press shilling publications, was slowly losing ground in middle-class culture; publishers of cheap serialized publications, conversely, were starting to realize its potential for thrill and sensationalism.[20] Between the 1830s and 1840s, publishers Edward Lloyd and G.W.M. Reynolds flooded the literary market with cheap serialized literature, creating a market that was parallel to, and distinct from, the middle-class one, and whose products were 'anathema to the established booksellers'.[21] As Greenwood's essay examined above shows, the word 'anathema' also describes very well how the middle class perceived the two publishers themselves. Lloyd and Reynolds were, first and foremost, two businessmen, and therefore the middle class openly doubted the genuineness of their support of the working class's self-improvement aspirations.[22] Actually, although only Reynolds pursued an active political career, a radical message transpired from their publications, which covered a wide range of topics and a variety of forms, and addressed working- and lower middle-class readers of both sexes and all ages[23]; significantly, they included fiction. Unsurprisingly, they had a greater appeal for the working-class public than the publications of the various societies for moral improvement had, even though they were sold while tracts were distributed for free. Literature produced by Lloyd and Reynolds could be purchased for the small sum of 1*d*, including the lurid penny blood series.

Penny bloods were the definition of cheap: churned out by underpaid hack-writers, they were either issued in penny miscellanies or printed separately on cheap paper at the low cost that characterized the serialized publications market. They were read aloud, and passed around until the paper fell to pieces, which explains the scantiness of original material available to scholars.[24] Penny bloods were, quite literally, read to destruction, thus fulfilling their only function, that is: satiating the craving for fiction of as many readers as possible. Their plots suited the modest requests of working-class readers: they were exciting, easy to read, and graphic, and they soon crystallized in a formula involving murder, betrayal, gender-shifting, and the occasional supernatural event (not to mention scantily clad damsels in distress). Their literary antecedents were disgraceful, from a middle-class perspective, as they readapted features and techniques from the Gothic novel and the broadsheet and shared several narrative features with melodrama.[25] Unrefined though they might appear, bloods were compatible with the generally more violent and graphic concept of entertainment that was popular among lower class individuals, despite the efforts of the middle and upper classes to change their ways.[26] Therefore, what distinguished Lloyd, Reynolds, and the penny blood authors whose biographies scholars have been able to reconstruct, was that they understood the needs of their target audience as much as the religious tract societies failed to understand them, and this was because they shared, or at least made an effort to comprehend, their audience's background.

Edward Lloyd may have died the equivalent of a multimillionaire, but he was born in the working class. He received little education, worked as a clerk, and attended the Mechanics' Institute,[27] after which, still a teen-ager, he started his own business in Shoreditch.[28] Haywood notes that, having moved his first steps in the literary business during the unstamped wars in the 1830s, Lloyd realized 'that populist publishing techniques were not incompatible with radical politics'.[29] In 1836, he published the *Lives of the Most Notorious Highwaymen* and *The Calendar of Horror* series,[30] reviving the *Newgate Calendar* tradition.[31] In the 1840s, he started the golden age of the penny blood genre by bringing on the market its two most prolific authors: in 1840–1841, Lloyd published Thomas Peckett Prest's *Ela, the Outcast*[32]; in 1843, *Ada, the Betrayed* by James Malcolm Rymer appeared on *Lloyd's Penny Weekly Miscellany*.[33]

Both authorship classification and mere biographical research in the penny blood genre are complicated by anonymity, the habit of having a different writer continuing the series if the original author was unavailable,[34] and the sometimes-arbitrary attributions made by scholars, as in Montague Summers's 1940 *Gothic Bibliography*. Scholars, though, have been able to reconstruct the main facts in the lives of Prest and Rymer. Prest, born around 1810, started his career writing adaptations of French farce and melodrama.[35] He then wrote imitations of Dickens for Lloyd, and later started writing his own fiction. Although he was one of the most prolific penny blood authors, he died a pauper at forty-nine, in 1859.[36] James ascribed Prest's ill-fortune, firstly, to the nerve-wracking rhythm of penny blood production, which impacted on his already weak health, and secondly, to Prest's inability to adapt his work to the changing tastes of his audience.[37] The penny blood market was merciless: authors who did not sell were dismissed.

Rymer's story is different. In his biography in the *Oxford Dictionary of National Biography*, Louis James writes that Rymer was born in 1814 in London (Holborn), his parents of Scottish origin, and that, primarily, he was interested in mechanics: records from 1841 register him as a civil engineer. There are grounds to suppose he attended the London Mechanics' Institute, and James suggested that Rymer and Lloyd may have met there.[38] Rymer's output was prodigious: he was rumoured to have been writing at some point as many as ten series simultaneously,[39] which he usually published under several anagrammatic pseudonyms, such as 'Malcolm J. Errym', or 'Malcolm J. Merry'.[40] This reticence to have his name directly connected to his penny bloods may be explained with his wish to write something more 'respectable' at some point. Indeed, James explains that Rymer, middle-class born, aspired to middle-class writing: from both his fiction and non-fiction transpires a certain contempt towards the ignorant masses, and his non-fictional writings bespeak 'a competent essayist in the style of Leigh Hunt'.[41] Yet, it was the fiction he wrote for the masses that made him famous and relatively wealthy (that is, for a penny blood author). In his essay 'Popular Writing', he wrote that '[i]f an author [...] wishes to become popular in the sense in which we use the term, that is, to be read by the majority, [...] he should [...] study well the animals for whom he is about to cater'.[42] In his biography of Rymer, James suggests that this ability to attune his writing to his audience's preferences originated in Rymer's work as a civil engineer, which brought him in touch with his targeted audience.[43] His sensitivity to his readers' tastes led him in the 1850s to move from Lloyd to John Dicks, whose publications were more refined: in 1858, *Reynolds's Miscellany*, published by Dicks, featured a tale by 'M.J. Errym'.[44]

We find trace of the change in taste that led Rymer to change publisher in Henry Mayhew's *London Labour and the London Poor*. In the section titled 'The Literature of the Costermongers', a costermonger interviewed about the literary tastes of his fellow professionals declared that '[w]hat they love best to listen to – and, indeed, what they are most eager for – are Reynolds's periodicals, especially the "Mysteries of the Court"'.[45] The man added: '[t]hey've got tired of Lloyd's blood-stained stories [...] and I'm satisfied that, of all London, Reynolds is the most popular man among them. They stuck to him in Trafalgar-square, and would again'.[46] This comment refers to the event that marked the beginning of Reynolds's Chartist political activity. In 1848, he made a speech at the radical gathering in Trafalgar Square, which was held to show support towards the Paris rebellion that ultimately led to the overthrow of King Louis Philippe. The chairman, Charles Cochrane, never showed up and the police were about to disperse the meeting, but Reynolds 'impulsively scrambled onto the platform' and his speech so heated the audience that he was followed home and continued it from the balcony of his own house.[47] Unlike Lloyd, Reynolds was born in the higher echelons of society, his father a senior naval officer and his mother the daughter of a Royal Navy captain. However, he showed no inclination to follow in his father's steps and left early from the Royal Military Academy at Sandhurst, where he had been entered at thirteen.[48] He espoused radical ideas early in his life, which brought him to Paris in the 1830s, where the 1830 Revolution consolidated his radical beliefs and introduced him to the idea of a 'literature without social boundaries'.[49] When he had to return, penniless, to his home country in the late 1830s, he started his career as a writer to support his family, which led him eventually to start his masterpiece, *The Mysteries of London* series, in 1844.[50] The first four volumes were published by Holywell Street publisher George Vickers; in 1848, they quarrelled, and Reynolds continued the series until 1856 with his own publisher John Dicks, slightly modifying the title to *The Mysteries of the Court of London*.[51]

The combination of unorthodox authors and publishers, radicalism, and humble or 'unsavoury' literary antecedents made the penny blood a subversive genre that contributed to shaping working-class alternative culture, as opposed to the mainstream culture of the middle and upper class. Penny bloods provided fiction for a class of readers that was seeking social, political, and personal improvement, and was inclined to do so without the guidance of its social superiors. Perhaps, the most incriminating feature of penny bloods was that they made obvious to the middle class that working-class tastes regarding fiction were beyond their control.

Whether more or less overtly political, the subversiveness of the penny bloods emerged, among other things, in the fact that several series tackled issues of social inequality relevant to their readership. Within the context of social and political ferment of the 1830s, there was a law that contributed to the climate of social injustice, perhaps in deceptively more subtle ways than the Reform Bill did: the 1832 Anatomy Act impacted heavily on the lives of the working class, or rather, on their deaths.

2 Bodies, Snatchers, and Doctors: The Path to the Anatomy Act

Since the early years of the nineteenth century, the medical fraternity had been seeking to legitimize its professional position and simultaneously to acquire political power. To do so, it was imperative for them to solve what they perceived to be two crucial issues in medical education management, namely, resurrectionists and the lack of dissection material: in the late 1820s, teaching human anatomy became a political concern.

As mentioned in the Preface, in his witness before the Select Committee for Anatomy, A.B. alludes to a 'prejudice' laypeople supposedly entertained against anatomy; this prejudice concerned the connection between dissection and murder sanctioned by the English law. Until the Eighteenth century, English anatomists had been studying the structure of the human body using the corpses of hanged felons,[52] which made the anatomist 'an executioner of the law'.[53] Then, in 1752, an Act of Parliament allowed judges to add dissection to the murderers' death sentence, in order to 'better Preve[nt] the Horrid Crime of Murder',[54] which explicitly connected the assassin to the lancet. The true impact of this change on the relationship between the poor and the medical world emerged with the population increase in mid-eighteenth-century England, which made it necessary to improve the medical system. Simultaneously, surgeons sought to become a profession in their own right, splitting up the Company of Barber-Surgeons in 1745; by 1800, they had become the Royal College of Surgeons, and placed increasing importance on the knowledge of anatomy.[55] Soon, the bodies of murderers became insufficient to meet the demand for corpses. As in the case of cheap literature, a new market rose to meet the new demand.

Bodysnatching was a thriving business, which occupied an ambiguous position before the law (technically, resurrecting a body was not theft, as a corpse did not 'constitute property')[56] and was unambiguously resented by the wider community: bodysnatchers profited from consigning the bodies of the innocent to the ignominy of the anatomist's scalpel, which, since 1752, was associated with the taint of murder. Reactions to the discovery of bodysnatching activity could verge on riot, and armed cemetery patrols were formed to prevent it. The patrols impacted on the body market: the increased risk of personal injury (or even death) for bodysnatchers boosted the price of the 'commodity', with a detrimental effect on the already tense relationship between resurrectioni sts and anatomists. From 1800, surgeons began discussing alternative options for sources of bodies and anatomy became a matter of public and political discussion.

The debate revolved around the conflicting religious, philosophical, and popular beliefs about the soul,[57] and was complicated by alarming rumours about the indecent way anatomist s treated corpses.[58] The situation deteriorated in the 1820s, when a further decrease in the body supply coincided with an increased demand triggered by the Royal College of Surgeons' manoeuvres to monopolize anatomy teaching.[59] Prices for bodies sky-rocketed, and the fees for a qualification in surgery became so heavy, while yet not granting the possibility to perform regular dissections, that students migrated to France to pursue their studies.[60] Finally, in March 1828, two members of the medical profession were prosecuted for bodysnatching,[61] setting a dangerous precedent. It was at this point that the Benthamites in Parliament took action.

Utilitarian philosopher and politician Jeremy Bentham was reputedly interested in the issue of anatomy studies.[62] Ruth Richardson recovered a series of letters from 1826 that show Bentham trying to pitch to Robert Peel a first draft of the Anatomy Bill. He proposed that all hospital patients should be made to make bequest of their bodies for dissection in case of death, which, since only the poor became hospital patients, implied he believed that paupers were the logical source of anatomy material.[63] Similar plans had been suggested repeatedly in the previous decade. John Abernethy had first made such a suggestion in his *Hunterian Oration* in 1816.[64] Also, Thomas Southwood Smith's article 'The Use of the Dead to the Living' first published in the *Westminster Review* in 1824, proposed using the bodies of paupers who died in hospital and workhouses, as French anatomy schools did. Peel, however, declined Bentham's suggestion on that occasion. After March 1828, though, the Benthamites took more decisive action and pushed for the creation of a Select Committee on Anatomy.[65] The Committee was 'a fine illustration of the working of Parliamentary Benthamism'[66]: its members were mostly Benthamites, the witnesses were partial to the solution proposed by Benthamites, and they ignored discordant voices.[67] The man who penned the *Report from the Select Committee on Anatomy* was Henry Warburton MP, the man who would give his name to the Anatomy Bill.[68] The Committee's agenda fundamentally envisaged the realization of Bentham's solution. The Report suggested, predictably, the use of the bodies of those who had been 'maintained at the public charge' in life and had died 'in workhouses, hospitals and other charitable institutions', in cases where they were not 'claimed by next of kin within a certain time after death'.[69] This was the first time the notion of 'not claimed', or 'unclaimed' (which appears one line below), one of the core concepts of the Anatomy Act, was mentioned.

Richardson notes that no discussion took place to clarify what exactly either 'unclaimed' or 'next of kin' meant.[70] In accordance with Bentham's leading doctrine of 'the greatest happiness for the greatest number',[71] the Report stated that this solution was necessary to guarantee protection to the rest of the community.[72] Besides, the selection it entailed would hurt no one, the Report asserted, as it was assumed that an unclaimed body meant 'indifference on the subject of dissection' on the part of the relatives.[73] This assumption did not consider that claiming a body meant paying for the funeral, which would have been impossible for some of the poor.[74] The Report dismissed accusations of class unfairness stating that the 'inconveniences' presented 'must be compared with those of the existing system' (i.e. bodysnatching), which after all already affected the poor, and repeated that 'where there are no relations to suffer distress, there can be no inequality of suffering, and consequently no unfairness'.[75] Yet, as Richardson notes, the use of distress as the measure for injustice, paired with the blurred definition of 'next of kin', disregarded any bonds of friendship that the poor might have formed outside the family circle,[76] ultimately denying the poor the status of community.

The first Anatomy Bill, presented in 1829 and dubbed the 'Midnight Bill' in the *Lancet* because of the secrecy that characterized its discussion in Parliament, did not pass.[77] While opposition to it had been inconsistent,[78] the Bill was resolutely rejected by the Lords: some members of the aristocracy advocated the 'right of the poor to a decent burial' and supported their aversion to be subjected to a practice that bore the mark of murder.[79] Others, most notably the Duke of Wellington, perceived the potential risk such a law represented for the already delicate political situation.[80] Things changed dramatically the following October, when, with perfect timing, certain gruesome accidents revealed that dealing in corpses had become so remunerative that some would literally kill to sell a body.

In November 1827, a lonely man died in rented lodgings in West Port, Edinburgh. His landlady's husband, Irish immigrant William Hare, and his compatriot William Burke, brought the body to Surgeon's Square: they had heard doctors paid well for corpses. Their quest for a purveyor led them to the private anatomy school of Dr. Robert Knox, who paid £ 7 10*s* for the body.[81] Obtaining such a sum over a corpse must have seemed a portent to the two lower class men. The second body they sold to Dr. Knox had not died from natural causes, nor did the following fourteen, between 1828 and 1829. Nobody suspected foul play. Burke and Hare had perfected a system of dispatching their victims that was 'practically undetectable until the era of modern forensics': they suffocated them, compressing the chest while simultaneously covering the mouth and nose, usually after intoxicating the victim with liquor.[82] Success, and the steady flow of money, made Burke and Hare greedy, which ultimately made the scheme fall apart. They murdered an Irishwoman named Margaret Docherty in Burke's lodgings while Ellen M'Dougal, Burke's companion, was hosting relatives. Although they had been sent away with an excuse while the homicide was being committed, M'Dougal's relatives eventually found Docherty's body, and alerted the police. When the police arrived, Burke and Hare had already removed the body to Surgeons' Square; yet, their incoherent justifications raised suspicions. Docherty's body was tracked to Dr. Knox's premises the following day. Hare turned King's evidence against Burke, who was found guilty and hanged on January 28, and his remains were publicly dissected. During the trial, the killing system was named 'burking', after Burke, the alleged inventor, and the wave of panic that ensued was consequently named 'burkophobia'.[83] Although the Edinburgh homicides were shocking, they had no substantial effect on the proceeding of the Anatomy Act through Parliament. Two years later, though, burking was discovered in London.

In November 1831, known bodysnatchers James May and John Bishop and occasional resurrectionist Thomas Williams laid the corpse of a teenage boy before Richard Partridge, professor of anatomy at King's College. The corpse was so strange that it even aroused the dissection-room porter's suspicions: it looked as if it had never been buried; the gums were toothless and, more alarmingly, bleeding. After a brief inspection, Partridge affected wanting coins to pay the men and went for the police.[84] The corpse was identified as Carlo Ferrari, an Italian boy, and the trial that ensued found the men guilty of the murder of two more victims: Fanny Pigburn, a Shoreditch woman, and a homeless boy named Cunningham. May was later acquitted, while Bishop and Williams were hanged on December 5, 1831, and their bodies were dissected. Their killing method classified as 'burking', although instead of suffocating their victims after dazing them with drink and laudanum, they would thrust them head first into the well in Bishop's backyard, adjacent to Williams's, in the East End neighbourhood of Nova Scotia Gardens, Bethnal Green. The case triggered another wave of burkophobia, which Warburton expertly exploited. Twelve days after Bishop and Williams were executed, he introduced his second Anatomy Bill.[85] During the Parliamentary discussions that ensued, whenever the argument of the opposition became too convincing, Warburton and his colleagues had but to remind the assembly about the 'late "enormities"'.[86] On the first of August 1832, the Anatomy Act was voted in.

3 Twisted Words: The 1832 Anatomy Act

The Anatomy Act indeed opened by mentioning the murders. The first fourteen lines of the document stated:

> '"Whereas a knowledge of the causes and nature of sundry diseases which affect the body, and of the best methods of treating and curing such diseases, and of healing and repairing divers wounds and injuries to which the human frame is liable, cannot be acquired without the aid of anatomical examination: and whereas the legal supply of human bodies for such anatomical examination is insufficient fully to provide the means of such knowledge: and whereas, in order further to supply human bodies for such purposes, divers great and grievous crimes have been committed, and lately murder, for the single object of selling for such purposes the bodies of the persons so murdered: and whereas therefore it is highly expedient to give protection, under certain regulations, to the study and practice of anatomy, and to prevent, as far as may be, such great and grievous crimes and murder as aforesaid:' BE IT THEREFORE ENACTED [...]"'[87]

This repetitive opening formula is deceptively simple. As Elizabeth T. Hurren observes, the twenty-one clauses constituting the definitive version of the Act were virtually unintelligible for the semi-literate poor, among whom oral culture still prevailed.[88] Indeed, it is likely that the law, or rather its summary hanging on workhouse walls, church doors, and in other spaces in which the pauper might gather,[89] was read aloud by the more skilled readers to the others. Hurren notes that the sonnet-like form of the first 14 lines, which would have sounded familiar to the pauper, disappears in the rest of the text, which is characterized by technical terms and complicated phrasing, while punctuation was scarce.[90] Article 2 gives an idea of the challenge the reading of the Act might constitute for an unskilled reader or a semi-literate listener:

> It shall be lawful for his majesty's said principal secretary of state or chief secretary, as the case may be, immediately on the passing of this act, or as soon thereafter as may be necessary, to appoint respectively not fewer than three persons to be inspectors of places where anatomy is carried on, and at any time after such first appointment to appoint, if they shall see fit, one

> or more other person or persons to be an inspector or inspectors as aforesaid; and every such inspector shall continue in office for one year, or until he be removed by the said secretary of state or chief secretary, as the case may be, or until some other person shall be appointed in his place; and as often as any inspector appointed as aforesaid shall die, or shall be removed from his said office, or shall refuse or become unable to act, it shall be lawful for the said secretary of state or chief secretary, as the case may be, to appoint another person to be inspector in his room.[91]

The endless repetitions ('appointment to appoint', 'one or more other person or persons to be an inspector or inspectors'), the countless 'aforesaid' and 'such', and the long, convoluted periods interspersed with technical terms clash with the clarity and rhythm of the opening section. The document becomes increasingly intricate as it proceeds to discuss the rights of the deceased and their relatives. Art. 7, for instance, separates the role of 'lawful owner' of the body from the relatives:

> It shall be lawful for any executor or other party having lawful possession of the body of any deceased person, and not being an undertaker or other party intrusted [sic.] with the body for the purpose of interment, to permit the body of such deceased person to undergo anatomical examination, unless, to the knowledge of such executor or other party, such person shall have expressed his desire, either in writing at any time during his life, or verbally in the presence of two or more witnesses during the illness whereof he died, that his body after death might not undergo such examination, or unless the surviving husband or wife, or any known relative of the deceased person, shall require the body to be interred without such examination.[92]

The article recognized the ownership of the body, and therefore decisional right over it, to a figure such as a workhouse master and, while it ratified his position—it is 'lawful' for him to have 'lawful' possession of the body and 'permit' the examination—it did not ratify that of the spouse or relatives. They could only 'require' that dissection should not happen, not 'lawfully prevent' or 'lawfully deny permission for' it. A further controversial point is that the dying person must be aware that they had to make explicit request not to undergo dissection, either in writing, which put the semi-literate poor in a disadvantaged position, or verbally in front of witnesses, which tied the execution of the dying pauper's will to the integrity of workhouse staff.[93] Finally, as with the

Report from the Select Committee, the expression 'known relatives' constituted a loophole. Richardson observes that the Bill did not state explicitly that it was the workhouse staff's duty to inform the relatives of an inmate's death, which made it plausible that 'relatives' absence could be construed as 'indifference''.[94] Art. 8 specified that the 'nearest known relative' was 'the deceased person's surviving husband or wife, or nearest known relative, or any one or more of such person's nearest known relatives, being of kin in the same degree'.[95] Therefore, the Act, as the Select Committee, failed to acknowledge the wider 'network' of friendship relationships noted by Richardson.[96] Notably, articles 10 and 14, which put the medical fraternity on safe ground, were surprisingly clear. Art. 10 established that licensed anatomist s could dissect 'the body of any person deceased', provided they had been authorized by the 'lawful' owner; art. 14 stated that a licensed anatomist could not be prosecuted for receiving, holding, and dissecting a human body, regardless of any irregularity committed by the lawful owner who authorized the dissection.

Art. 18 stated that violating the Act qualified as 'misdemeanour', an offence of lesser degree, which was punished with either three months imprisonment or a £ 50 fine. This light penalty contrasts with the condemnation of burking stated in the opening, and effectively failed to change the situation since 1813, when bodysnatcher Joseph Naples, found in possession of a stolen body, brazenly asked the court if this was 'a bailable offence' and, receiving a positive answer, he 'made a bow and retired'.[97] Richardson and Hurren observe two further noteworthy details. Art. 16 repealed the 1752 Act by which dissection became distinctive of murder sentences. Richardson, though conceding that this represented a step forward, suggests that it is unlikely the pauper would find 'much comfort in the knowledge that [they] would be dissected on the slab *instead* of a murderer, rather than *alongside* one'.[98] By contrast, she continues, the Act never mentioned the body trade (basically failing to eliminate 'the known motive for both bodysnatching and burking'); in a subtle way, she suggests, this rather constituted a ratification of the body trade.[99] This situation remained unchanged in subsequent legislation regulating the exercise of the medical profession: in 1858, the medical fraternity's quest for professional legitimization produced the Medical Act, which named and grouped the institutions in the kingdom that were authorized to release licenses to practitioners, and the organs and people appointed to control the courses and register licensed practitioners.

Simultaneously, it made achieving a license in one of the 'official' medical schools the only legal way to practise medicine—which meant, among other things, that attending courses at private anatomy schools would no longer provide a valid license. While ostensibly aimed to prevent the spreading of quacks and unskilled practitioners, this law left the Anatomy Act's provisions unvaried while concentrating medical education (and medical students' fees) to a restricted number of institutions, to which flocked increasingly high numbers of trainees. As the century progressed, the Anatomy Act continued to constitute the basis of later regulations, which were characterized by greater and greater emphasis on dissection, such as the 1885 Medical Act extension; this 'cumulative impact', as Hurren terms it, influenced the demand for dissection material across the century.[100]

Furthermore, Hurren notes that the clause that repealed the dissection of murderers is also the only instance in which the word 'dissection' appears in the text.[101] In the rest of the Act, it is referred to as 'Anatomical Examination', or simply 'Examination', which, Hurren observes, implied 'touch[ing], prob[ing], inspect[ing], and view[ing]' rather than systematic dismemberment.[102] This euphemistic phrasing created the impression that the text of the Act was retaining information from the reader/listener.[103] In actual fact, the 1832 Anatomy Act was a law that, in theory, promoted the development of medical science while simultaneously setting in place safety measures; in practice, it was an exercise in rhetoric, against which the pauper—semi-literate, socially powerless, and politically underrepresented—could not possibly win.

4 The Poor, the Doctor, and the Penny Blood: A New Analytical Perspective

In 1963, Louis James's groundbreaking *Fiction for the Working Man* first outlined the history of cheap serialized fiction, which included the penny bloods. He opened the preface to his work stating that the book was 'primarily the study of *literature*',[104] acknowledging this status to working-class fiction for the first time. Since James's 1963 seminal work, research on the penny blood and penny dreadful has progressed, although scholarship remains scanty, compared to other Victorian genres. The late-1980s–early-1990s were a particularly productive moment, during which John Springhall and Anne Humpherys produced the bulk of scholarship. I must mention also Elizabeth James and Helen R. Smith's *Penny Dreadfuls and Boy's Adventures*, the catalogue of the Barry Ono Collection of Victorian cheap serialized fiction at the British Library (1998). Its importance cannot be overestimated, as it guides the scholar through the single largest surviving collection of original material in the field. The early 2000s focused on re-evaluating the figures of Lloyd and Reynolds, most notably in Ian Haywood's *The Revolution in Popular Fiction* (2004) and in Humpherys and James's 2008 essay collection *G.W.M. Reynolds: Nineteenth-Century Fiction, Politics, and the Press*. Stylistic was instead the focus of Helen R. Smith's 2002 study *New Light on Sweeney Todd, Thomas Peckett Prest, James Malcolm Rymer and Elizabeth Caroline Grey*. This study provided crucial new data on the authorship of Rymer's and Prest's works, including convincing evidence that Rymer, and not Prest, was the author of *The String of Pearls*, or *Sweeney Todd, the Demon Barber of Fleet Street* (henceforth *Sweeney Todd*).

Simultaneously, scholars started exploring the single narratives; most notably, Sally Powell's 2004 essay 'Black Markets and Cadaverous Pies: The Corpse, Urban Trade, and Industrial Consumption in the Penny Blood' and Robert L. Mack's 2007 study *The Wonderful and Surprising History of Sweeney Todd* focused on Rymer's narrative of cannibalistic pies. Both scholars highlight the connection between this narrative and the Anatomy Act: Powell argued that the anxieties about the body trade the industrial revolution sparked in the working class, which the Anatomy Act reinforced, were 'most pertinent to the working-class reader', who felt the threat the anatomist posed to their bodily integrity.[105] Mack also highlighted this connection, noting that bodysnatching, burking, and the body trade fuelled a set of anxious associations between anatomy and cannibalism in the popular mind.[106] The Demon Barber is also at the core of Rosalind Crone's analysis of the penny bloods within the broader landscape of working-class 'violent' entertainment in *Violent Victorians* (2010). Notably, Crone expands Powell's argument connecting *Sweeney Todd* and the Anatomy Act by addressing the overcrowded conditions of the mid-century metropolis. Crone reads Todd's 'murder-machine' as 'a macabre solution to the mounting dead in the city', 'designed to deal with the urban masses and ensure that their remains would be put to good use' in a way that resembled the efficient yet soulless logic of the Anatomy Act.[107]

This brief survey of the scholarship produced on the penny blood highlights the conspicuous absence of an extensive systematic study of a selection of specimens focusing on what the bloods can yield as narratives. The approaches that emerge from the survey above—production and distribution history, history of the publishers and publishing environments, studies on authorship and isolated narratives—are invaluable tools to understanding the literary and historical context in which the genre developed. However, they do not allow us to notice recurrence of features, themes, and repetitive patterns. Following the methodological approach adopted by Powell, Mack, and Crone, with this book, I propose to fill this gap by examining a cross section of penny bloods within their cultural and literary context. More specifically, my examination finds its focal point in a trait that Powell, Mack, and Crone include in their exegeses of *Sweeney Todd*: the connection between the pauper, their fiction, and the Anatomy Act. To do so, I delve deeper in the relationship between paupers and medicine, examining the scholarship on the history of the Anatomy Act.

Ruth Richardson's 1987 study *Death, Dissection, and the Destitute* revolutionized studies in medical history, providing the first revisionist reading of the Anatomy Act. Whereas it had long been regarded solely as a landmark in the progress of medical science, Richardson considered the Anatomy Act from the point of view of its impact on the Victorian poor. In 2012, Elizabeth T. Hurren's study *Dying for Victorian Medicine* progressed research on this issue, reconstructing the history of various anatomy schools, in London and in the provinces, and of the body traffic between 1834 and 1929. Both Richardson and Hurren challenge the idea promoted by the supporters of the Act that paupers were not 'distressed' by its provisions. Recent studies on burking confirm this argument and extend the examination of the issue to its diffusion through the press. Sarah Wise's *The Italian Boy* (2004) and Lisa Rosner's *The Anatomy Murders* (2010), examining the London and Edinburgh burkers' cases, respectively, illustrate how burking affected the most exposed members of the lower class and how the two cases became part of the sensational life of the country through press coverage and fictional readaptation. Wise, most notably, connects the London burkers and the penny blood genre, making a convincing case for the two Bethnal Green murderers having inspired Reynolds's monstrous Bethnal Green resident, the Resurrection Man. In this book, I consider the information these historical studies provide on the context of the Anatomy Act and compare it with the way medicine and the dissected/dismembered corpse are represented in the narratives examined. Recent studies on the novel from mid to late nineteenth century adopt a similar medical historical approach. Tabitha Sparks' study *The Doctor in the Victorian Novel* (2009) uses the figure of the doctor as a lens to examine marriage as the core element of the Victorian novel plot, looking at the progressive power shift from fiction to medical science alongside contemporary efforts the medical fraternity made to 'locate the knowledge of human life in physiology rather than literary subjectivity'.[108] In 2013, Pamela Gilbert's essay 'Sensation fiction and the medical context' focused on the medical characters of sensation novels in relation to the development of the mid-century figure of the clinician, arguing that the figure of the medical hero that developed in mid-century novels concretized in sensation fiction in a character made disquieting by the quasi-supernatural power of his clinical gaze.[109] Both Sparks and Gilbert focus on the impact of mid-century medical advancements on the novel, and examine how the increasing importance of medicine in Victorian society influenced the representation of the doctor in a range of representative narratives. In this book, I apply this paradigm to representations of doctors, dissection, and displaced/dismembered cadavers in a selection of penny bloods in the light of early- to mid-century changes in anatomy studies, to outline how the genre represented doctors and dissection after the passing of the Anatomy Act.

On a broader level, this book argues the importance of the penny bloods as marginalized narratives. I use the term 'marginalized' here with the double meaning of narratives that, traditionally, are considered low-brow, hence unworthy of scholarly interest, and of 'suppressed historical narratives of marginalized groups',[110] in this case the poor. As mentioned above, the research area of penny bloods is still comparatively unexplored, mainly due to two reasons: first, the long-lasting nineteenth-century stigma attached to the bloods, and second, the fact that penny blood narratives do not, from a qualitative perspective, match the standards of canonical fiction.[111] Since 1983, Franco Moretti challenged the exclusiveness implicit in the concept of 'canon', advocating the inclusion of 'mass literature' in literary studies to chart literary history 'with great theoretical precision and historical fidelity' and acquire new perspectives for reading the works that constitute the so-called canon.[112] With this study, I contribute to broaden the picture of Victorian literature, facilitating the departure from nineteenth- and twentieth-century views that still influence the perception of the Victorian penny blood, building a systematic, structured study of a set of penny bloods with respect to the context of the Anatomy Act and its impact on the readership of the genre. This 'cross section' paradigm shifts the focus from the examination of (one selected specimen of) the genre as part of the broader serialized publications context to examining the genre itself within its historical context, particularly relating to medical history; this approach allows to identify common traits, continuities, discontinuities, and evolutions that do not emerge from the case study or book history approach, and the medical history angle contextualizes one of the bloods' outstanding traits, the recurrent representation of the displaced/dismembered corpse. Instead of dismissing this trait as the fruit of the somewhat unrefined tastes of the bloods' readership, I consider it in the light of the momentous change that the passing of the Anatomy Act constituted for the people whose disposal of remains was determined by poverty, with the purpose of achieving a better understanding of the genre's role towards its readers.

The first challenge the penny blood genre poses to the scholar is the sheer size of the corpus. Series were started, abruptly interrupted if unsuccessful, or, conversely, stretched to impossible length if they sold well. This creates the second problem, that is, the tendency of the penny bloods to be bulky, rambling narratives. A study that proposes to analyse, to dissect, so to speak, this literary monster should first attempt to isolate a selection of specimens. The selection I used in this book includes three texts that can be reasonably considered key-narratives in the genre based on their success among the intended readership and durability of their fame: *Varney the Vampyre* and *Sweeney Todd* by J.M. Rymer and *The Mysteries of London* by G.W.M. Reynolds. The fourth narrative, *Manuscripts from the Diary of a Physician*, again ascribed to Rymer, stands out from the multitude of series collected by Barry Ono and preserved at the British Library because of its medical theme: it is possibly the only penny blood series to focus on the medical world and, furthermore, its representation of medical men displays major similarities with that of *Varney*. In the analysis, I consider first Rymer's three works, in chronological order, and then Reynolds's *Mysteries*. This structure allows an appreciation of the continuities in the works ascribed to Rymer, making it possible to compare the three narratives and then test their traits against Reynolds's work, highlighting common features and/or differences in the approach of the two authors. Allowing for the scarcity of original material, I chose the editions on which the analysis is developed considering their faithfulness to the original penny bloods, at least in their volume form. The edition of *Manuscripts* used here is a British Library Historical Collection reproduction of the original 1844 (series 1) and 1847 (series 2) volumes that Lloyd reprinted, allegedly, 'at the request of a large number of its readers'.[113] *Varney* was originally published in penny parts by Lloyd,[114] then reprinted in volume form, which the Dover edition I use here reproduces. Mack's 2007 edition of *Sweeney Todd* matches faithfully the copy at the British Library, a volume form of the original series issued in Lloyd's *The People's Periodical and Family Library* from November 1846 to March 1847.[115] Finally, the copy of *Mysteries* I used is a digitized reproduction of the original 1845 volume 1, originally published in penny parts by George Vickers starting 'in or around October 1844'.[116] In the four analytic chapters, I examine the key-features that emerge from a preliminary observation of the relationship between doctors and displaced/dismembered corpses in the narratives: this relationship exposes issues of monstrosity—the individual responsible for the displacement/dismemberment of a body always display monstrous traits—as well as issues of powers. Characters in positions of power, who usually coincide with doctors or people connected to the resurrection of bodies, are represented as more or less monstrous depending on the way they exercise their power. Ascent to, or fall from, power—power to act, to speak, to decide—is reflected in the constant vertical movement of bodies through space: cadavers being resurrected, or living people being plunged into the darkness below, cause the boundaries of life and death to become blurred and death and dissection to become uncomfortably close to the physical space of the living. My analysis puts these features in context with respect to the Anatomy Act, adopting a perspective that combines new historicism, spatial theory and discourse theory.

The four analytic chapters consider the texts alongside the narratives that can be evinced from newspaper articles, diverse types of reports, and, of course, the 1832 Anatomy Act. I compare the rhetoric of these sources to the language strategies the penny bloods deployed when discussing issues of death, dissection, and medicine. I treat the bloods as 'marginalized' narratives, meaning narratives that, traditionally, are considered low-brow, hence unworthy of scholarly interest, and are the 'suppressed historical narratives of marginalized groups',[117] in this case, the poor. In so doing, I deconstruct the '*master narrative*', that is, 'a narrative told from a single cultural point of view that, nevertheless, presumes to offer the only accurate version of history'[118] that still influences our perception of the penny blood. The issue I address here, therefore, is not whether the bloods are 'good' literature, but, more pragmatically, what role they performed towards their audience as a literary product. So far, studies performed on penny bloods as narratives and as commodities in the marketplace have been pointing in this direction. My purpose is that of advancing this process, examining the penny blood as an 1830s–1840s 'cultural artifact'[119] to ascertain what information it can provide about the 'marginalized narrative' of the relationship between their poor readership and anatomy, expanding our understanding of the penny bloods' role in the circulation of anatomy discourses.

While the discourse about medicine concerned the poor directly, a set of strategies were put in place to exclude them from the conversation. More or less overtly, the promoters of the Act wished the poor to know very little, and understand even less, about what the document implied for them. Adopting as a starting point Robert T. Tally's principle that '[i]n a manner of speaking, literature also functions as a form of mapping, [...] providing points of reference by which [readers] can orient themselves and understand the world in which they live',[120] I examine how the selected narratives mapped out anatomy and dissection for their readers, allowing them to understand a discourse from which they were excluded, representing and transforming Victorian London's 'geospace'. This term, coined by Barbara Piatti, stands for the 'real' space as opposed to the 'imaginary' space represented in the literary text, which will never be the same.[121] I consider the strategies penny bloods deployed to circulate the anatomy discourse through the representation and transformation of London's geospace in their narratives, specifically the mimesis of the geospace and the use of the narrative space as a metaphor. Particularly, the recurrent dichotomy of surface versus subterranean space is crucial to ascertain the role of the bloods as a channel for discourse circulation. Across the nineteenth century, the middle and upper class took a variety of measures to keep at bay their increasing anxiety about the underground space, which translated to concerns about the criminal underworld, poverty, and, ultimately, death. These measures, which spanned from the manipulation of the urban landscape to such laws as the Anatomy Act, were based on the power dynamics that regulated class distinction and therefore worked downwards through the social strata. Similarly, penny bloods represented the vertical movement of bodies through the social and geographical space as ultimately based on these power dynamics. In my analysis, I map out these representations, outlining how they translated for their audience the power dynamics that regulated their relationship with death and dissection and circulating discourse s of power associated with medicine and anatomy.

The Anatomy Act, as a social and historical phenomenon, was intrinsically part of a power discourse. As a written document, the Act is a practical example of the relation between power and knowledge. Controlling the circulation of information, establishing restrictions on what can be discussed, the control, in brief, over knowledge, is what characterizes power according to Foucault. He also stressed the nature of power as a relation and its being consequently subject to negotiation. The elusive language of the Act was meant to prevent the paupers' protests. However, this elusiveness problematizes the power relationship the Act entailed and the position of the poor within it, triggering a process of negotiation of power in which penny bloods had a role. From the point of view of Foucault's theory of power discourses, the narratives examined here worked as a channel for circulating an alternative discourse, in contrast with the dominant one about medicine and anatomy. Through their spatial representations, but even more subtly and effectively through the deployment of specific linguistic strategies, penny bloods allowed the poor a certain degree of power, that of discussing, understanding, and ultimately finding a way to cope with their position with respect to medicine and dissection.

The first chapter analyses *Manuscripts from the Diary of a Physician*, examining how the narrative represents the medical man, and the medical student, as a positive or negative character depending on his work ethic. The second chapter moves on to examine the same dichotomy of positive/negative medical man in *Varney the Vampyre*, in which the nature of the medical man is more directly related to issues of masculinity, nervous mind states, and the clinical gaze. What chiefly distinguishes the positive doctor from the negative in *Varney* is his application of the clinical gaze as the language of truth: the less truthful the language, the less trustworthy the doctor. Issues of truth also form the basis of the analysis of *Sweeney Todd* in the third chapter. Although the medical discourse is less obvious, it nonetheless works in the background of this narrative, in which control over the transmission of knowledge regulates power relations between characters. An intrinsic 'Londoner' quality and the representation of a trapdoor-paved urban landscape connect *Sweeney Todd* to Reynolds's *The Mysteries of London*, which is analysed in the fourth and last chapter. In Reynolds's narrative, the presence of an actual burker-bodysnatcher recreated for the reader one of the darkest pages of medical history, while moving the discussion of the power struggle between the pauper and anatomists to a more markedly political ground.

Notes

1. James Greenwood, 'A Short Way to Newgate,' in *The Wilds of London* (London: Chatto and Windus, 1874), 158.
2. Ibid.
3. From Greenwood's essay also emerges one of the core points of discussion regarding the genre, that is, the distinction between penny 'blood' and 'dreadful'. 'Penny dreadful' is often used as an umbrella-term for the whole genre. However, scholars generally agree in differentiating the 'penny bloods', the 1830–1860 production, gorier and more violent, and addressing a public of mixed ages, from the 'penny dreadful', the post-1860 production, less violent and aimed specifically to a juvenile audience. See James Malcolm Rymer, *Varney the Vampyre; or: The Feast of Blood*, ed. Curtis Herr, 2008th ed. (Crestline, CA: Zittaw Press, n.d.), 10–13. In this book, I use purposefully the term 'penny blood' as I examine works that belong to the first category.
4. Greenwood, 'A Short Way to Newgate,' 159.
5. Ibid., 166.
6. Ibid., 166–69.
7. Rosalind Crone, *Violent Victorians—Popular Entertainment in Nineteenth Century London* (Manchester: Manchester University Press, 2012), 2–6.
8. Richard D. Altick, *The English Common Reader—A Social History of the Mass Reading Public, 1800–1900* (Chicago and London: The University of Chicago Press, 1957), 141–43.
9. Graham Law and Robert L. Patten, 'The Serial Revolution,' in *The Cambridge History of the Book in Britain*, ed. David McKitterick (Cambridge: Cambridge University Press, 2009), 144.
10. Simon Eliot, 'From Few and Expensive to Many and Cheap: The British Book Market 1800–1890,' in *A Companion to the History of the Book*, ed. Simon Eliot and Jonathan Rose (Oxford: Blackwell, 2007), 293.
11. Ian Haywood defines the stratagem of placing the part publications at a low price that was nonetheless beyond working class reach as 'typical of bourgeois compromise'. Ian Haywood, *The Revolution in Popular Literature* (Cambridge: Cambridge University Press, 2004), 163.

12. Louis James, *Fiction for the Working Man 1830–1850—A Study of the Literature Produced for the Working Classes in Early Victorian Urban England* (London: Oxford University Press, 1963), 115; and Altick, *The English Common Reader*, 103–4.
13. See Altick, *The English Common Reader*, 104–6.
14. Ibid., 106.
15. Charles Dickens, *Bleak House*, ed. Stephen Charles Gill (Oxford: Oxford University Press, 2008), 122.
16. James, *Fiction for the Working Man*, 120–21. James explains that fiction, from an Evangelical's viewpoint, was dangerously close to lying (because it was 'not true') and was therefore 'a sin'. Ibid., 120.
17. See Altick, *The English Common Reader*, 130–37.
18. Crone, *Violent Victorians*, 167.
19. Eliot, 'From Few and Expensive to Many and Cheap,' 295.
20. James, *Fiction for the Working Man*, 72–87; and Crone, *Violent Victorians*, 176–77.
21. Graham Law, *Serializing Fiction in the Victorian Press* (Basingstoke: Palgrave, 2000), 20.
22. The question: 'Fathers of the cheap press or "able speculators"?' posed (and answered) by Haywood succinctly summarises the still ongoing debate around these two controversial radical publishers. Haywood, *The Revolution in Popular Literature*, 162.
23. Ibid., 140.
24. Even when they remained intact, penny bloods and dreadfuls were not preserved by vendors or libraries.
25. Crone, *Violent Victorians*, 173–78.
26. Crone deconstructs the whole concept of 'taming' of working-class culture by middle-class culture, arguing that the wide circulation of tales such as *Sweeney Todd* suggests that they were not a 'deviation from the norm of respectability', but a substantial part of working-class culture that continued to exist, resisting the middle-class push towards a more 'tamed' system. Ibid., 2–6.
27. Ibid., 169.
28. John Sutherland, *The Longman Companion to Victorian Fiction*, 2nd ed. (Harrow: Longman, 2009), 382.
29. Haywood, *The Revolution in Popular Literature*, 163.
30. Sutherland, *The Longman Companion to Victorian Fiction*, 382.
31. Crone, *Violent Victorians*, 173.
32. The series was originally published in 104 parts by J.K. Edwards. Elizabeth James and Helen R. Smith, *Penny Dreadfuls and Boy's Adventures—The Barry Ono Collection of Victorian Popular Literature in The British Library* (London: The British Library Board, 1998), 75.

33. Ibid., 83.
34. '[I].e. drunk, pregnant, or dead'. Rymer, *Varney the Vampyre; or: The Feast of Blood*, 16.
35. Sutherland, *The Longman Companion to Victorian Fiction*, 517.
36. Ibid.
37. James, *Fiction for the Working Man*, 33, 96.
38. Ibid., 36.
39. Ibid., 33.
40. Sutherland, *The Longman Companion to Victorian Fiction*, 214.
41. James, *Fiction for the Working Man*, 37.
42. James Malcolm Rymer, 'Popular Writing,' *Queen's Magazine: A Monthly Miscellany of Literature and Art* 1 (1842): 172.
43. Louis James, 'Rymer, James Malcolm (1814–1884),' *Oxford Dictionary of National Biography*, 2004, http://www.oxforddnb.com/view/article/53817.
44. Ibid.
45. Henry Mayhew, *London Labour and the London Poor* (London: Griffin, Bohn, and Company, 1861), 25.
46. Ibid.
47. Anne Humpherys and Louis James, eds., *G. W. M. Reynolds—Nineteenth-Century Fiction, Politics, and the Press* (Aldershot: Ashgate, 2008), 5.
48. Ibid., 1.
49. Ibid., 2.
50. Ibid., 4.
51. Ibid.
52. Ruth Richardson, *Death, Dissection and the Destitute* (Chicago and London: The University of Chicago Press, 1987), 32.
53. Ibid., 34.
54. Ibid., 35–36.
55. Elizabeth T. Hurren, *Dying for Victorian Medicine—English Anatomy and Its Trade in the Dead Poor, c. 1834–1929* (London: Palgrave Macmillan, 2012), 78–80.
56. Richardson, *Death, Dissection and the Destitute*, 58, 70–71.
57. Ibid., 93.
58. Ibid., 95–99.
59. Ibid., 101.
60. Ibid., 101–2.
61. Ibid., 107.
62. Ibid., 108.
63. Ibid., 110.
64. Ibid., 108.

65. Ibid., 107.
66. Ibid., 109.
67. Ibid., 108.
68. Ibid.
69. 'Report from the Select Committee for Anatomy' (London, 1828), 9.
70. Richardson, *Death, Dissection and the Destitute*, 123–24.
71. Ibid., 108.
72. 'Report from the Select Committee for Anatomy,' 10.
73. Ibid.
74. Richardson, *Death, Dissection and the Destitute*, 124.
75. 'Report from the Select Committee for Anatomy,' 10.
76. Richardson, *Death, Dissection and the Destitute*, 124.
77. Ibid., 157.
78. On pages 152–57, Richardson explains that divisions within the radical faction regarding the role of science made their opposition to the Bill ineffective, and ultimately led to its passing. This point will be explored in greater detail in the next chapters.
79. Ibid.
80. Ibid., 157–58.
81. Lisa Rosner, *The Anatomy Murders* (Philadelphia: Universty of Pennsylvania Press, 2010), 28.
82. Ibid., 54. While the Caledonian Mercury alleged that Burke and Hare must have been instructed by an expert in human anatomy, Burke declared that, by lying over the chest of his victims, he had no intention of compressing the lungs, but merely of pinning them down. See ibid., 55–56.
83. Richardson, *Death, Dissection and the Destitute*, 194.
84. Sarah Wise, *The Italian Boy—Murder and Grave-Robbing in 1830s London*, Pimlico 20 (London: Jonathan Cape, 2004), 23.
85. Ibid., 248.
86. Richardson, *Death, Dissection and the Destitute*, 198.
87. *Anatomy Act* (London: H.M.S.O., 1832), 902. Emphasis in the original.
88. Hurren, *Dying for Victorian Medicine*, 21–22.
89. Ibid., 21.
90. Ibid., 27.
91. *Anatomy Act*, 903.
92. Ibid., 903–4.
93. See Richardson, *Death, Dissection and the Destitute*, 207.
94. Ibid.
95. *Anatomy Act*, 904.
96. Richardson, *Death, Dissection and the Destitute*, 124.

97. 'Police—Hatton Garden,' *The Times*, 1813, 2, https://doi.org/10.1007/s13398-014-0173-7.2.
98. Richardson, *Death, Dissection and the Destitute*, 207, Richardson's emphasis.
99. Ibid., 208.
100. Hurren, *Dying for Victorian Medicine*, 4.
101. Ibid., 29.
102. Ibid., 28.
103. Ibid.
104. James, *Fiction for the Working Man*, xi, James's emphasis.
105. Sally Powell, 'Black Markets and Cadaverous Pies: The Corpse, Urban Trade and Industrial Consumption in the Penny Blood,' in *Victorian Crime, Madness and Sensation*, ed. Andrew Maunder and Grace Moore (London: Ashgate, 2004), 46.
106. Robert L. Mack, *The Wonderful and Surprising History of Sweeney Todd—The Life and Times of an Urban Legend* (London: Continuum, 2007), 41–43.
107. Crone, *Violent Victorians*, 189.
108. Tabitha Sparks, *The Doctor in the Victorian Novel* (Farnham: Ashgate, 2009), 7–8.
109. Pamela K. Gilbert, 'Sensation Fiction and the Medical Context,' in *The Cambridge Companion to Sensation Fiction*, ed. Andrew Mangham (Cambridge: Cambridge University Press, 2013), 187.
110. Lois Tyson, *Critical Theory Today* (London and New York: Routledge, 2006), 287–88.
111. The term 'canonical' refers to a concept first theorized by F.R. Leavis in 1948 as the 'novelists in English [that are] worth reading'; this definition was famously restrictive, as Leavis limited it to 'Jane Austen, George Eliot, Henry James, and Joseph Conrad – to stop for the moment at that comparatively safe point in history'. Frank Raymond Leavis, *The Great Tradition—George Eliot, Henry James, Joseph Conrad* (London: Chatto and Windus, 1948), 9. While Leavis's restrictive conception has long since been abandoned, the term is still used to indicate, in a way, fiction whose standards are 'acceptable' and by the quality of which all other fiction is measured and deemed more or less 'acceptable'.
112. Franco Moretti, 'The Soul and the Harpy,' in *Signs Taken for Wonders: On the Sociology of Literary Forms* (London: Verso, 1983), 15, 16.
113. James and Smith, *Penny Dreadfuls and Boy's Adventures*, 87.
114. Law and Patten, 'The Serial Revolution,' 153.
115. James and Smith, *Penny Dreadfuls and Boy's Adventures*, 79.

116. Humpherys and James, *G. W. M. Reynolds—Nineteenth-Century Fiction, Politics, and the Press*, xvii.
117. Tyson, *Critical Theory Today*, 287.
118. Ibid.
119. Ibid., 286.
120. Robert T. Jr Tally, *Spatiality* (London: Routledge, 2013), 2.
121. Barbara Piatti, *Die Geographie Der Literatur: Schaulplätze, Handlungsräume, Ramuphantasien* (Göttingen: Wallstein Verlag, 2008), 32–33. Quoted in ibid., 52.

List of Works Cited

Anatomy Act. London: H.M.S.O., 1832.

Altick, Richard D. *The English Common Reader—A Social History of the Mass Reading Public, 1800–1900*. Chicago and London: The University of Chicago Press, 1957.

Crone, Rosalind. *Violent Victorians—Popular Entertainment in Nineteenth Century London*. Manchester: Manchester University Press, 2012.

Dickens, Charles. *Bleak House*. Edited by Stephen Charles Gill. Oxford: Oxford University Press, 2008.

Eliot, Simon. 'From Few and Expensive to Many and Cheap: The British Book Market 1800–1890.' In *A Companion to the History of the Book*, edited by Simon Eliot and Jonathan Rose. Oxford: Blackwell, 2007.

Gilbert, Pamela K. 'Sensation Fiction and the Medical Context.' In *The Cambridge Companion to Sensation Fiction*, edited by Andrew Mangham, 182–95. Cambridge: Cambridge University Press, 2013.

Greenwood, James. 'A Short Way to Newgate.' In *The Wilds of London*. London: Chatto and Windus, 1874.

Haywood, Ian. *The Revolution in Popular Literature*. Cambridge: Cambridge University Press, 2004.

Humpherys, Anne, and Louis James, eds. *G. W. M. Reynolds—Nineteenth-Century Fiction, Politics, and the Press*. Aldershot: Ashgate, 2008.

Hurren, Elizabeth T. *Dying for Victorian Medicine—English Anatomy and Its Trade in the Dead Poor, c. 1834–1929*. London: Palgrave Macmillan, 2012.

James, Elizabeth, and Helen R. Smith. *Penny Dreadfuls and Boy's Adventures—The Barry Ono Collection of Victorian Popular Literature in the British Library*. London: The British Library Board, 1998.

James, Louis. 2008. 'Rymer, James Malcolm [Pseuds. M.J. Errym, Malcolm J. Merry] (1814–1884), Novelist and Journal Editor.' *Oxford Dictionary of National Biography*. November 16, 2015. http://www.oxforddnb.com/view/10.1093/ref:odnb/9780198614128.001.0001/odnb-9780198614128-e-53817.

James, Louis. *Fiction for the Working Man 1830–1850—A Study of the Literature Produced for the Working Classes in Early Victorian Urban England*. London: Oxford University Press, 1963.

Law, Graham. *Serializing Fiction in the Victorian Press*. Basingstoke: Palgrave, 2000.

Law, Graham, and Robert L. Patten, 'The Serial Revolution.' In *The Cambridge History of the Book in Britain*, edited by David McKitterick. Cambridge: Cambridge University Press, 2009.

Leavis, Frank Raymond. *The Great Tradition—George Eliot, Henry James, Joseph Conrad*. London: Chatto and Windus, 1948.

Mack, Robert L. *The Wonderful and Surprising History of Sweeney Todd—The Life and Times of an Urban Legend*. London: Continuum, 2007.

Mayhew, Henry. *London Labour and the London Poor*. London: Griffin, Bohn, and Company, 1861.

Moretti, Franco. 'The Soul and the Harpy.' In *Signs Taken for Wonders: On the Sociology of Literary Forms*, 1–41. London: Verso, 1983.

'Police—Hatton Garden.' *The Times*. September 10, 1813, 2, https://doi.org/10.1007/s13398-014-0173-7.2.

Powell, Sally. 'Black Markets and Cadaverous Pies: The Corpse, Urban Trade and Industrial Consumption in the Penny Blood.' In *Victorian Crime, Madness and Sensation*, edited by Andrew Maunder and Grace Moore, 45–58. London: Ashgate, 2004.

'Report from the Select Committee for Anatomy.' London, 1828.

Richardson, Ruth. *Death, Dissection and the Destitute*. Chicago and London: The University of Chicago Press, 1987.

Rosner, Lisa. *The Anatomy Murders*. Philadelphia: University of Pennsylvania Press, 2010.

Rymer, James Malcolm. 'Popular Writing'. *Queen's Magazine: A Monthly Miscellany of Literature and Art* 1 (1842): 99–103.

Rymer, James Malcolm. *Varney the Vampyre; or: The Feast of Blood*. Edited by Curtis Herr. Crestline, CA: Zittaw Press, 2008.

Sparks, Tabitha. *The Doctor in the Victorian Novel*. Farnham: Ashgate, 2009.

Sutherland, John. *The Longman Companion to Victorian Fiction*. 2nd ed. Harrow: Longman, 2009.

Tally, Robert T., Jr. *Spatiality*. London: Routledge, 2013.

Tyson, Lois. *Critical Theory Today*. London and New York: Routledge, 2006.

Wise, Sarah. *The Italian Boy—Murder and Grave-Robbing in 1830s London*. Pimlico 20. London: Jonathan Cape, 2004.

CHAPTER 2

Manuscripts from the Diary of a Physician: Power, Ethics, and the Super-Doctor

Unless the keen scholar spots *Manuscripts from the Diary of a Physician* (hereafter *Manuscripts*) while sifting one of the few available penny blood catalogues, it is unlikely that they will ever hear of it. The series, first appeared in *Lloyd's Penny Sunday Times* in the early-to-mid 1840s, is virtually unknown nowadays.[1] Yet, the Barry Ono collection includes a two-volume reprint, which in the penny blood world is a testimony of great success: the series must have attracted quite a vast number of faithful readers, if Lloyd embarked in the production of a volume edition, more expensive than the serialized version.

This narrative stands out in the flock of penny bloods starting from its very own title: *Manuscripts from the Diary of a Physician*. The wording itself is peculiar and denotes a theme and narrative structure distinctive in the genre: these are 'manuscripts' from a 'diary', a redundant phrasing that sounds as if the author is taking great pain to convince the reader that this text contains authentic information. While it was not uncommon in penny bloods that characters should recover lost diaries and letters, *Manuscripts*, through its title, constructed itself as a found document in its own right, turning the reader into the finder. Moreover, it is hardly possible to miss the similarity between this title and that of a more famous, more highbrow series issued on *Blackwood's Magazine* between 1830 and 1837: *Passages from the Manuscripts of a Late Physician*, by Samuel Warren. *Passages* was a work of fiction where an unnamed physician narrated some of his most interesting cases; the

A. Gasperini, *Nineteenth Century Popular Fiction, Medicine and Anatomy*, Palgrave Studies in Literature, Science and Medicine,
https://doi.org/10.1007/978-3-030-10916-5_2

stories celebrated middle-class values and virtues, in line with *Blackwood's* audience's expectations, and indeed Sparks notes that the patients' dramatic stories often backgrounded the role of the physician-narrator himself.[2] The *Blackwood's* series was so realistic that it sparked outrage among surgeons, who believed it a gross violation of patient-doctor confidentiality.[3] *Manuscripts* engaged in a similar attempt, possibly with the purpose of imitating Warren's narrative, and perhaps ride on its success—not an unusual practice in the cheap serialized press industry.[4] The similarities between the *Blackwood's* series and the *Lloyd's Penny Sunday Times*', though, do not stretch much further than the title and the narrative device of the anonymous storytelling doctor; this is chiefly due to the fact that *Manuscripts* catered to a different audience than *Blackwood's* own, and this emerges clearly in how each series addresses the theme of dissection.

In the previous chapter, I stated that the medical theme makes *Manuscript* possibly unique in the penny blood genre. Penny blood readers loved reading the adventures of criminals and murderers, compared to which the life of a physician might look quite uneventful. What could his diary have contained that would have interested the average action-seeking, gore-loving penny blood reader? The answer is in the title of the opening narrative of the series: 'The Dead Restored; or, The Young Student'. From the first page, the physician narrator and the whole series are linked to the most gruesome and terrifying aspect of medicine: the sight of, and the contact with, the dead body. *Manuscripts* is in fact an excellent example of how penny bloods could deploy the sensation and gore that came to define them as a 'low' genre to discuss controversial themes that concerned their readership. While encompassing a variety of situations, the series engages from the onset with the least savoury aspects of medical studies, and with the ethical debate around poverty, power, and dissection. This point constitutes a crucial distinction between *Passages* and *Manuscripts*, which I will explore in more detail: in Warren, the bodysnatching episode is a playful scene that aims to create complicity between the fictional physician and its middle-class reader. Its narrator-protagonist brings to the foreground the challenge posed by relatives objecting to dissection of their dead loved ones and portrays the bodysnatching doctor as almost heroic. *Manuscripts*' working-class audience, by contrast, was offered a much murkier story that, while ultimately absolving the unnamed physician, problematized the doctors' attitude towards lower class people's bodily integrity. *Manuscripts*' discussion of these themes, and the (possibly intentional) resonance with Warren's *Passages*, demonstrate that, far from being solely a gore- and profit-oriented narrative detached from its own literary context, it participated in a contemporary literary trend: the rise of the doctor as a fictional hero, adapting it to its themes and to its audience's tastes.

Gilbert notes the rise to popularity, around mid-century, of the figure of the medical hero.[5] While she refers, specifically, to the development of the phenomenon in sensation fiction, this genre stems in some measure from other forms of popular print culture in the earlier part of the century[6]; these, as *Manuscripts* demonstrates, included the penny bloods. The figure of the physician as author-narrator of a penny blood is consistent with the phenomenon Gilbert describes, as physicians were potentially melodramatic figures. They were the 'gentlemen of the medical world',[7] people entrusted with family secrets. Hence, the narrator's profession promised salacious, scandalous, or horrifying details, the kind of titillation the readers of cheap serialized fiction looked for when purchasing literature. From this perspective, the narrative's construction as a fragment of reality had, first, the commercial function to lure the reader into purchasing the series: the title enticingly hinted at the secret nature of the content through the word 'diary', simultaneously exploiting a figure that was gathering prominence in popular literature, that of the medical man. The series' Introduction further capitalized on the readers' voyeurism: *Manuscripts*' fictional narrator-physician explains his reader that he will relate some of 'those 'moving incidents' [...] which are unfolded to none more freely than to the physician', but he assures that he 'will only speak of the past' to preserve his living patients' privacy, therefore portraying himself as a trustworthy, and most of all real, physician. Then, he proceeds to introduce the first narrative, granting that it 'proves that truth is stranger, far stranger, than fiction'.[8]

Manuscripts' configuration of its narrative as the diary of a medical man, then, tapping into fashionable literary trends in popular fiction, aimed to fulfil the chief objective of penny bloods according to the editors, namely, profit. Here, however, I consider the series as an early example of the rise of the medical man as a fiction hero; it is therefore necessary to examine the relationship between the thematic freedom afforded to the 'lowlier' penny blood genre and the medical-historical context in which the narrative appeared. Sparks states that, until the reputation of medical men improved towards mid-century, the figure of the doctor was not considered novel material also due to its unsavoury connection with bodysnatching.[9] These elements, instead, suited the penny blood genre perfectly and, as it appears from the contents of the first episode, *Manuscripts* exploits them fully. Indeed, the simultaneous presence of a student and a corpse in the first episode's title would trigger in the mind of the mid-Victorian reader several connections to the issue, still debated in the 1840s, of the use of paupers' bodies as dissecting material for training anatomy students. In this chapter, I explore how *Manuscripts* discusses, elaborates, and problematizes this topic, focusing on two specific episodes: the above-mentioned episode one of series one, 'The Dead Restored; or The Young Student' (hereafter 'The Dead Restored'), and episode seven of series two, 'The Long Subject; or, The Unexpected Denouement' (hereafter 'The Long Subject'). These two episodes stand out from the considerable number constituting the long series, and are simultaneously related to each other, because they share a few crucial features: they both revolve around unscrupulous doctors exhuming, or attempting to exhume, a pauper's body for dissection purposes. Furthermore, both episodes pivot on cases of premature burial, in which the physician plays a critical role in restoring the health of the person who was buried alive.

The little, but crucial, authorial information that we possess on this series is pivotal to some of the issues I address here. As with most penny bloods, *Manuscripts* was published anonymously. Montague Summers' *Gothic Bibliography*, reported in Louis James' *Fiction for the Working Man*, ascribes the authorship to James Malcolm Rymer and so did Helen R. Smith in 2002 in *New Lights on Sweeney Todd*. Although, as we have seen, when it comes to penny bloods and penny dreadfuls, authorship is a delicate matter,[10] I agree with Smith's conclusions, since *Varney the Vampyre* and *Sweeney Todd*, both ascribed to Rymer, display recurrent similarities to *Manuscripts*' style. Significantly, the use of medical terms and the construction of medical characters in *Manuscripts* and in *Varney* are remarkably similar. I would therefore venture that it is possible to say, with a fair amount of certainty, that Rymer is *Manuscripts*' author. This provides interesting points for analysis: his Scottish origins offer some ground for reflection on the Edinburgh setting of the very first narrative of the series, 'The Dead Restored'. Moreover, Rymer's profession as a civil engineer, as James suggests, may have put him in contact with the environment of the Mechanics' Institute and the idea of 'useful knowledge'; this offers an interesting perspective to analyse the recurrence of Utilitarian vocabulary and philosophical ideas in the two narratives.

In 'The Dead Restored', the author of the diary, the unnamed physician, relates an adventure he had as a student in Edinburgh. The protagonist is walking through the poorest neighbourhood of the city with his friend and colleague Musgrove, when, upon hearing laments coming from one of the wretched houses, they discover the desperate Major Sinclair, weeping over the corpse of his beautiful little daughter Mary. Mary's corpse, and her inexplicable death, become an obsession for Musgrove, who decides to disinter her and dissect her. His determination to carry out a dissection borders on madness, or, as he terms it, 'monomania'. It is worth noting that Gilbert identifies monomania as a core element of sensation fiction and as a trait characterizing also Dickens's 1860s' fiction.[11] Alongside the theme of the medical hero, therefore, *Manuscripts* embraced also the interest in altered mental states shown by related narrative forms. The (soon to be) physician is not keen on Musgrove's project, but decides to help him anyway, to prevent him from asking for help from someone less trustworthy, thus compromising himself. However, when they reach the cemetery on an appropriately stormy night, they find the distraught father weeping over the daughter's tomb, and the sight makes Musgrove renounce his plans. To their horror, however, they see Major Sinclair has disinterred the corpse of the child with the purpose of burying it in her mother's tomb by the river. The two medical students approach the deranged man to make him desist, and it seems everything will eventually turn out well. However, Musgrove suddenly realizes that the child is sitting upright with her eyes open, watching them. The sight kills Major Sinclair on the spot. It is eventually discovered that his daughter was alive and well and had been buried while in a state of coma.

The events of 'The Dead Restored' provide ground for examining the use of space in the narrative. From the description of the locations in the story emerges an effort to reproduce the geospace of the city of Edinburgh in the narrative space. By comparing a map of early-nineteenth-century Edinburgh with these descriptions, it is possible to find some of the landmarks that characterize the narrative space. The two medical students find the Sinclairs in a slum and, later, Mary is buried at Caulknows, '[t]o the left of the Leith road'.[12] In Victorian Edinburgh, there were slums around Princes Street, near the spot where Waverley train station is now. Heading east, Princes Street makes a turn into Leith

Street, which conceivably could be the 'road to Leith' near which the narrative locates the cemetery where Mary Sinclair is to be buried. Since no evidence of such place as 'Caulknows' can be found in historical maps, the space is probably fictional (Fig. 1).

However, next to the 'road to Leith' there is Calton Hill with Calton burial ground, the name of which somewhat resembles that of the fictional Caulknows. While the Edinburgh setting could be ascribed, in part, to Rymer's Scottish descent, the nature of the narrative as a penny blood suggests other reasons connected to the history of crime and medicine, which at times were tightly interlaced, of that city. The story obviously tries to connect the snatching of Mary Sinclair from her grave with the past of bodysnatching of Edinburgh. Likewise, Musgrove's determination to dissect Mary's body is redolent of the long-lasting debate around the ethics—or lack thereof—of the behaviour of the anatomists, which saw the Edinburgh medical community at the centre of debate on various occasions.[13]

Resurrection happens also in 'The Long Subject', in which the physician is now a seasoned practitioner and lives in a village in the south of England. 'Long' Hannibal Jeffries, the idle and drunkard, but famously tall, husband of the doctor's landlady, dies of exposure on the night in which Mrs. Jeffries, exasperated by his disorderly conduct, refuses him admittance to the house. The woman is distraught and blames herself for her husband's death, although her whole community tries to console her—except a certain Revd. Gluck, a preacher, who

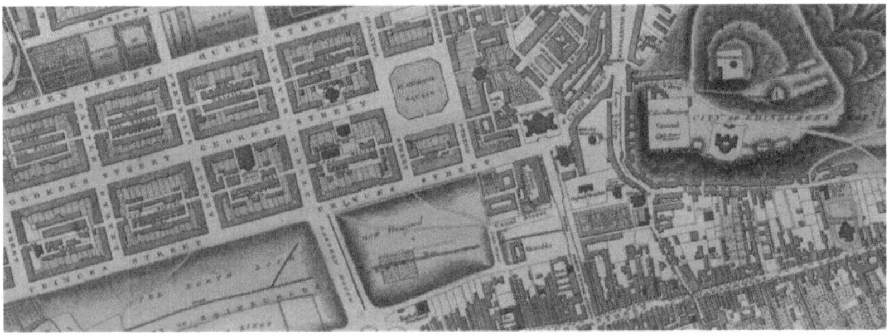

Fig. 1 Old and New Town of Edinburgh and Leith with the proposed docks—detail. Edinburgh: John Ainslie, 1804. Reproduced with the permission of the National Library of Scotland

asserts that she is doomed. Meanwhile Garratt, a colleague and friend of the doctor, is founding a Literary and Scientific Institution, and would like to have the skeleton of Long Hannibal as a specimen for its museum, considering his exceptional body frame. Garratt asks the physician to help him disinter the body, which he condescends to do, on condition that everything must be done so that the widow shall suspect nothing. On the night that follows the burial, Garratt's party digs up the body; as they are refilling the tomb, the physician watches the body, which starts moving. He thus discovers that Long Hannibal Jeffries had been thrown in a coma by exposure and is actually alive. This is convenient, for the night he (apparently) died, he had just received the news that an uncle of his had passed in the West Indies, leaving him four thousand pounds. His brother Tom had brought him the news; Hannibal meant to give him a thousand pounds, but Tom, who became entitled to the whole sum upon his brother's (apparent) death, would have shared none with the widow and children. Long Hannibal is reunited with his wife, through a tactful tale the doctor tells her (in which he manages to put the grave-robbing operation in the best possible light) and the cruel brother Tom is chased out of the house. The apparent death is explained by a prank played on Tom to test his generosity. The adventure cures Hannibal of his penchant for drinking and he and his family move to London, where they lead a respectable life.

'The Long Subject' engages more openly with the ethical implications of resurrectionism for dissection purposes than 'The Dead Restored' does. The emphasis on the 'usefulness' of the 'knowledge' that can be derived from opening a corpse, which characterizes the whole narrative, bears remarkable similarity to the language of Utilitarian philosophy. It suggests an attempt to justify surgeons who purchased bodies stolen from graves, or at least to come to terms with the idea that certain bodies did end up on the dissecting table. The absence of professional resurrectionists in both 'The Long Subject' and in 'The Dead Restored' is thus explained. The narratives mean to show the doctors in the best possible light, even when they exhume bodies, in a frame within which the ends justify the means.

Although the sensationalism necessary to guarantee the success of a penny blood is certainly the central element of these narratives, they arguably display an intention to discuss the ethical implications of the act of bodysnatching. They both tap into the public's dislike for dissection and resurrectionism, as emerges from the murky settings and the presence of arrogant, monstrous doctors. By contrast, they stand out among other popular forms of discussion of these issues, such as cartoons, by proposing a positive representation of anatomy and medical men. The discourse on anatomy and dissection developed in the narratives is built on a dialectic structure that juxtaposes rationality and irrationality, or insanity, hubris and humanity, science and superstition.

1 Monomaniac Monsters and Balanced Heroes: The Medical Man Reimagined

Monsters in *Manuscripts* are less overtly 'monstrous' than in the other penny bloods this book examines. This is a series that is (purportedly) based on the diary of a medical man who, as the reader gradually discovers, takes pride in his rationality; therefore, there are no supernaturally monstrous creatures, but rather monstrous humans, who are deceitful, murderous, and mad. In the two selected episodes, monstrosity is embodied by a medical man. The extant relationship between madness and doctors in the two narratives is rooted in the socio-historical context in which the series appeared. The 1830s and 1840s saw Victorian society expressing concerns about the balance of members of the medical fraternity, especially in relation to anatomy studies and dissection.[14] These concerns proceeded from the combination of the secrecy that characterized medical training, the medical fraternity believing that laypeople would not understand their methods and challenge them, with the disquieting intelligence that leaked from the closed doors of the dissection room into the pages of periodicals.

Truly, studying medicine in the nineteenth century was a rough matter, a powerfully physical experience that required the surgeon-to-be to make several adjustments at a psychological level, or to withdraw from the course. Although the discipline was advertised as a contemplative experience for a private club of well-dressed gentlemen, in the actual dissection room, students found themselves up to their elbows in gore, their nostrils full of the smell of decaying flesh, working on unappealing pieces of dismembered corpses for hours on end. Even the sexual dimension of the experience was drastically different from the expectations built in prospective students: while the advertising material suggested that dissection mostly entailed examining the bodies of beautiful females, the corpses in the dissection room were chiefly male.[15] Hurren notes the public's fear about the effects that this experience might have on students, leading them to develop unhealthy interests towards the patients and even perform experiments on them.[16] Dissection did, in fact, sever the cultural ties that bound the medical student to perceiving the dead body as an object of reverence and his fellow human beings as such, and not as patients. The avowed purpose of dissection, as William Hunter enunciated it around 1780,[17] was that of 'inform[ing] the *head*, guid[ing] the *hand*, and familiariz[ing] the *heart* to a kind of necessary inhumanity'.[18] The concept of 'necessary inhumanity', which is defined today as 'clinical detachment',[19] was then new and disconcerting. It implied that anatomy required medical students to relinquish their moral and religious values and start acting contrarily to what was classified as human and Christian behaviour. From the surgeon's perspective, students learned to operate with a steady hand; from the uninitiated's perspective, they learned to contemplate with detachment the idea of mutilating other human beings and of touching and dismembering corpses. They became, in a word, monstrous.

This perceived monstrosity stemmed also from another concept that developed in medical sciences in the nineteenth century, namely, the medical gaze. In *The Birth of the Clinic*, Foucault discusses the supremacy of the gaze in nineteenth-century medical practice, arguing that during the (comparatively) brief life span of nineteenth-century medicine and its focus on pathological anatomy, '[t]he 'glance' ha[d] simply to exercise its right of origin over truth'.[20] The clinician's gaze, therefore, established what truth was, and it was not the gaze of 'any observer', but that 'of a doctor supported and justified by an institution [...] endowed with the power of decision and intervention'.[21] The public felt the medical gaze as a manifestation of power acutely. Firstly, it represented the power of the trained clinician to individuate illness, the malfunctioning of human anatomy as only the proficient anatomist could know it. This power transformed the patient from unique individual into a set of 'constitutional and disease characteristics' subject to the gaze of the clinician.[22] Secondly, it represented the power to look, to pry into what was normally concealed from view. The intimate observation of a naked, open body that surgeons-to-be and anatomists had the power to perform disquieted laypeople, especially if that body was female. Even more disquieting was the anatomists' apparent impatience to exercise their gaze, to the point of starting a business relationship with individuals who were ready to desecrate a tomb to make a sale. The clinical gaze exercised in the dissection room and the hospital thus became a prying, voyeuristic gaze in the mind of the public, the exercise of power over the ultimate powerless entity, the diseased or dead body.

Manuscripts is profoundly rooted in this context, notwithstanding its humble literary origins, and possibly because of them. The working-class audience of the penny bloods was a group that more than any other was subject to the power of the medical gaze: they were observed and touched by the surgeons-in-training when receiving free treatment, and their bodies were the chief source of subjects for dissection, before and after the Anatomy Act. The themes of madness, voyeurism, and dissection emerge powerfully in the medical characters of the two narratives examined. In 'The Dead Restored', the narrative juxtaposes two types of medical students: a mad one, who is prone to monstrous behaviour and to abusing his power, and a balanced one (the fictional author as a student), who does not yield to monstrosity. This latter figure is connected to the mature, balanced physician of 'The Long Subject', who combines excellent medical performance with prowess and mettle. Through the young and mature physicians, the narrative builds a positive re-invention of the medical man figure.

The character of the medical student Musgrove embodies Victorian concerns about the impact anatomy studies could have on the mind of students. From the first paragraph, Musgrove is characterized as someone who lacks sensitivity, but who prides in his status of medical student and feels superior to the rest of society because of it. When he and the protagonist hear the laments coming from one of the poor houses they are passing by, they have two markedly different reactions that reveal two different perceptions of the boundaries between themselves and the private sphere of the poor members of their community:

> 'Let us go in,' said [Musgrove].
>
> I hesitated a moment, for there was ever to me something sacred about grief which repressed all curiosity.
>
> 'Pho! pho!' said my young companion, 'our profession is our passport to such scenes. We may be of some service.'[23]

The protagonist shrinks from imposing his presence on a stranger in grief, whereas Musgrove is ostentatiously dismissive of such scruples. He treats his status of medical trainee as a 'passport' that allows him to break boundaries that people are usually expected to respect, such as private property, or the intimate moment of grief. Although his suggestion that they 'may be of service' makes one suppose that perhaps he means well, his nonchalance in intruding into the house and life of the pauper in distress suggests the urge to indulge his own voyeuristic curiosity. He perceives the pauper as an object to gaze upon, and himself in the position of the observer, free to perform a close inspection that engages his eye, but not his feelings. The narrative reinforces this impression by outlining a distant, dismissive medical community in the background through Major Sinclair's reaction when he sees the two medical students. Believing them to be the doctors who left his pleas for help unanswered, he cries: 'Fiends! fiends! [...] In my poverty and bitter destitution [...] I sent for ye, and ye came not'.[24] Although the two students explain that they are not the surgeons to whom Sinclair appealed and even leave a small contribution to the funeral, the covert text is that the Edinburgh medical community is too proud, and too interested in money, to provide treatment for the destitute. Musgrove's character is framed as the offshoot of such medical community, which engages with the destitute only insofar they constitute interesting subjects. However, in Major Sinclair's house, the medical student finds interesting material for his gaze in the lifeless body of Mary Sinclair.

The reader surveys Mary's body through the young physician's gaze, who describes her as

> a young girl of apparently seven or eight years of age. The beauty even of that face in death exceeded anything I had ever beheld. There was still a faint tinge of colour on the cheeks, the beautiful lips were slightly apart, and through the long silken eyelashes a glimpse of clear blue eyes could be obtained: a flood of golden hair hung over her breast and the pillow on which she lay.
> I was quite fascinated by the still picture before me. She looked like some beautifully executed piece of wax work.[25]

Notably, the medical student's gaze registers the corpse as beautiful. The morbidity of this detail would not have escaped the expert eye of the average penny blood reader, who would have been looking for gory and disturbing passages. Additionally, a nineteenth-century reader would have connected the comparison between the corpse and a 'beautifully executed piece of waxwork' with the Anatomical Venus, wax models of beautiful females with long hair, which indeed seemed rather asleep than dead. These were used as anatomy models for students (wax internal organs could be found under the movable 'lid' of the torso), but they were also the central pieces of the anatomy exhibitions open to the public, and *Manuscripts*' readership was likely familiar with them.[26] The problematic element in the connection between Mary's corpse and a wax work, and hence a Venus, is the process of sexualization these figures underwent. Ludmilla Jordanova underscores the attempt at a 'realistic' representation of the female body in these models, which presented 'naturalistic colouring' and 'meticulous details such as eyelashes and eyebrows', and the markedly feminine and erotic elements that characterized this realism.[27] The Venus was always in a recumbent position on silk or velvet cushions, and her alluring pose sometimes recalled famous works of art, such as Bernini's Saint Teresa.[28] Her head was adorned with flowing human hair, and she wore a pearl necklace around her neck.[29] In his description, the young physician of *Manuscripts* focuses on Mary's 'flood of golden hair', her eyelashes, and 'tinge of the skin'. These details were the same ones that defined the wax model's realistic reproduction of the female body as a beautiful sight, as a perfect work.[30] Jordanova argues that the realism that characterized the Venus was 'uncanny' to behold[31]; in *Manuscripts*, however, the disturbing element is not so much the uncannily lively corpse as the onlooker, the student who registers the corpse as beautiful, even sensually so. By establishing a connection between the reader and popular Victorian displays such as these, the description of Mary's corpse through the young physician's eyes is a first hint in the narrative to the perception of the gaze of the medical student on the dead body as morbidly voyeuristic.

The corpse awakens the interest of both students; however, here too they have radically different reactions to the stimulus. While his description of the corpse suggests that anatomy training has altered the young physician's gaze and he now perceives a corpse as an attractive object, the sense of respect for the dead child that transpires from his words assuages the reader. The protagonist can draw the line between his medical gaze and his human, and humane, perception of the dead body. By contrast, Musgrove's perception of the corpse rapidly assumes markedly necrophiliac tones. He is suspiciously keen to learn the exact location where the dead girl will be buried. He confesses that he 'should like vastly to know what she died of' and he suggests that '[b]y getting – the body' they 'might know'.[32] While the physician is 'not much shocked at the proposition' because his 'professional education' had freed him from his 'prejudices', he 'shr[inks] from disturbing the repose of that young creature', and suggests that they '[l]eave her alone'.[33] Musgrove, however, is determined:

> [...] I must do it [...] the idea has haunted me, sleeping or waking [...] I believe I shall never know peace till I find out what that child died of.
> Why, Musgrove, are you mad?
> It almost amounts to a monomania [...] I am resolved, with you or without you, to possess myself of that girl's body tonight.[34]

Musgrove is obsessed with Mary's corpse, so much so that he terms it a 'monomania'. As he felt entitled to enter a pauper's house to satisfy his curiosity, he now feels entitled to examine the corpse of the dead destitute girl. From his perspective, the girl is already a 'subject', a body, which he can, and must, subject to his gaze. While the young physician observes the girl's body from a respectful distance, Musgrove is convinced that the pursuit of knowledge that characterizes his profession entitles him to disregard commonly accepted boundaries, including those imposed by the funerary rites. Believing himself, as a medical man endowed with power and allowed to make decisions, superior to social and religious conventions, he commits a sin of hubris that almost leads him to madness. The fact that he admits to a monomaniacal state adds strength to his disturbing, necrophiliac closing remark: 'I am resolved [...] to possess myself of that girl's body tonight'. The medical student thus becomes a mentally unstable character, reiterating Victorian concerns about the impact of anatomy training on students' minds and about the appropriateness of their gaze over the dead (female) body.

From this perspective, the gaze of the two medical students on Mary Sinclair's corpse, and consequently the *Manuscripts* series, belongs to a context in which the medical gaze on the corpse was being represented, and scrutinized, through different channels. Pictorial representations of dissection offered contrasting views on the all-powerful medical gaze, and while some celebrated it for the medical community, others focused on the anxieties about necrophilia and metaphorical cannibalism that it awakened in the wider public. Hurren examines the 1864 chalk drawing 'The dissection of a beautiful woman directed by J. CH. G. Lucae (1814–1855) in order to determine the ideal female proportions' (Fig. 2), which exemplifies the type of image that was used to advertise anatomy courses for students.[35] In the scene, three smartly dressed doctors examine—that is, they *observe*—a beautiful female corpse, not unlike

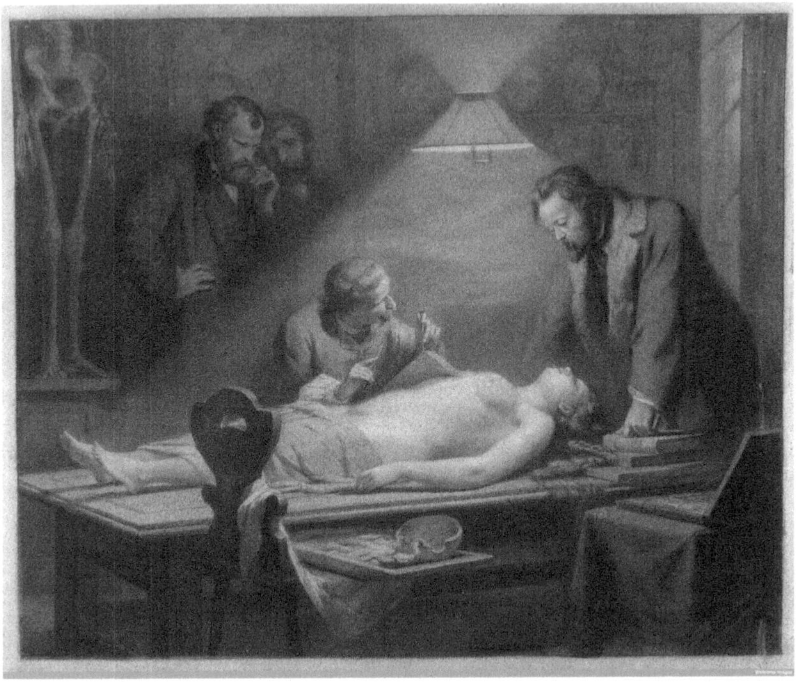

Fig. 2 Hasselhorst, Johann Heinrich. The dissection of a young, beautiful woman directed by J. Ch. G. Lucae (1814–1885) in order to determine the ideal female proportions. 1864. © Wellcome Library, London

an Anatomical Venus, with the purpose of determining the exact shape of female beauty.

As Hurren notes, Hasselhorst's representation of the proceedings of a dissection room is not realistic.[36] Blood and gore are conspicuously absent, while the scene exalts the clinician's power to access the inaccessible with his gaze. Jordanova observes that Lucae focused on the 'physical basis of female beauty', and argues that the image, particularly the detail of the sheet of skin lifted from the woman's breast, suggests that '[t]he corpse is indeed being undressed scientifically'.[37] The title itself reveals the true aim of the painter, which is not that of offering a realistic representation, but of celebrating the medical gaze. This bloodless dissection is performed on a 'beautiful' woman to allow a surgeon to 'determine' the parameters of female beauty: the image invites the anatomist-to-be to imagine himself as part of a select group whose trained gaze not only does engage in the delightful observation of naked female beauty, but also, being able to access what others cannot, produces authoritative knowledge that is endowed with the status of truth. Foucault states that '[t]he genesis of the manifestation of truth is also the genesis of the knowledge of truth'.[38] Since the clinician's gaze affords him a privileged access to the genesis of the manifestation of truth (in this case, the corpse), he enjoys direct access to the knowledge of truth. Consequently, his diagnosis, his words, acquire the status of truth, of knowledge. This kind of pictorial representation of dissection, therefore, advertised the power of accessing, scrutinizing, and evaluating as the privilege of the dissecting room's inhabitant, the same power Musgrove claims for himself as soon as he lays eyes on Mary Sinclair. In his mind, his ability to access the truth through his gaze is tantamount to his entitlement to do so whenever he wants.

By contrast, cartoons represented the gaze of the medical man and the medical student as grotesquely voyeuristic. This type of representation posed uncomfortable questions on the boundaries of the 'necessary inhumanity' advertised by modern medicine and its promotion of pathological anatomy. Discussing *The Lancett Club at a Thurtell Feast* (Fig. 3) by Thomas Rowlandson, which represented the dissection of the murderer John Thurtell, Flanders notes that the anatomist 'is grotesquely caricatured, while the corpse of Thurtell is entirely realistic'.[39] The title itself implied that the surgeon and his students, conceived as a sort of private club, fancied human flesh, and that Thurtell's dead body represented, for them, an appetizing feast. In the print, the surgeon dips his bare hands avidly into the open abdomen of the corpse, the fingers contracted in a clawed position, while a mass of gore and organs tumbles from the open body into a bucket. The medical students display equally caricatured facial expressions ranging from excitedly interested to cruel. Only one of them is touching the corpse, keeping one side of the abdomen open for his senior to perform his task; all the others stand and watch eagerly, hovering behind the surgeon's shoulders to get a better view of the corpse. The corpse, as Flanders notes in her comment on the plate, is 'almost classically handsome' and does not resemble the actual Thurtell as portrayed by Joseph Hunt during the trial.[40] Indeed, it could be any corpse. Its face has nothing of the contorted expression of Tom Nero in Hogarth's famous illustration of dissection *The Four Stages of Cruelty* (1751, Fig. 4). While Hogarth's print's title implies that

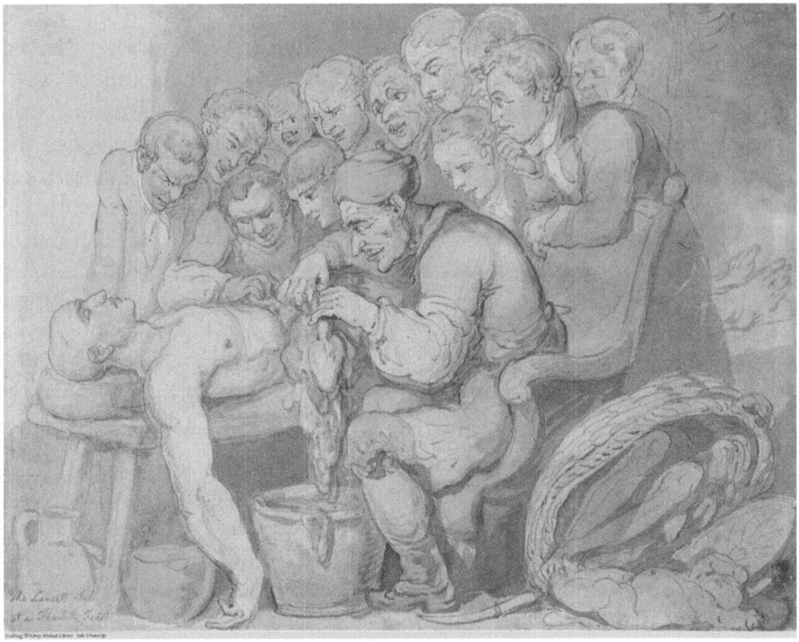

Fig. 3 Rowlandson, Thomas. The Lancett Club at a Thurtell Feast. © Yale University, Harvey Cushing/John Hay Whitney Medical Library

dissection is the just reward for Nero's crimes and the corpse is neither idealized, nor celebrated, *The Lancett Club* portrays the cadaver as a banquet for the voyeuristic surgeons. The form of the cartoon, belonging to a more popular context (meaning it was intended for the general public, as opposed to the specialist public addressed by Hasselhorst), represents the dissected body as powerless under the hands and gaze of the medical fraternity. This type of representation emerges in the way Mary Sinclair's body, which stands for an ideal and idealized beauty disconnected from the actual age of the girl, is powerless under Musgrove's monomaniacal, voyeuristic gaze. The representation of Musgrove as a threat to Mary's body therefore conforms to the tendency of popular culture to portray the clinical gaze as threatening.

Unlike the corpse in in Rowlandson's print, the idealized body in 'The Dead Restored' is female. This difference is unsurprising, as it addressed a specific concern of the public about anatomy studies. The

Fig. 4 Hogarth, William. Caricature of an anatomical dissection, depicting the cruelty of dissected criminal avenged by the heartlessness of the anatomists; 'The Reward of Cruelty'. 1751. © Wellcome Library, London

fact that the anatomist's gaze on the female corpse had explicit sexual connotations emerged from both pictorial and sculptural representations of the female corpse (as we have seen) and, more alarmingly, from rumours of grotesque jokes made behind the closed doors of the anatomy theatre. Anatomists and medical students were famously fond of jokes and pranks that disregarded the dignity of the dead body, especially the female one.[41] Musgrove's character in 'The Dead Restored' embodies Victorian concerns about disrespect towards the (female) corpse in the dissection room. He would remind the Victorian reader of the average medical student, or young anatomist, such as Richard Partridge, who Wise describes as the typical self-made man who 'quickly adopted the condescension and waggishness of the gentleman-born'.[42] Musgrove, with his dismissive 'pho! pho!' and his high opinion of his own medical person, tallies with this type of character, and his excessive, inappropriate interest in Mary Sinclair's body reiterates popular representations of the medical gaze as voyeuristic and necrophiliac.

The mad medical student of 'The Dead Restored' tapped into the popular perception of the medical student as a bogeyman. He is a monstrous figure, dehumanized by dissection and inebriated with the power of the medical gaze, which makes him disregard both the living and the dead. The narrative, however, suggests that his condition is not irreversible. When they are finally on the point of disinterring the girl, Musgrove and the young physician spot the father weeping on the tomb, and the young physician asks Musgrove if, after seeing this, he will still pursue his aim. He replies: 'No [...] I will not. My mind has strangely altered. I have lost the wish to do so'.[43] His answer sounds as if the sight of Major Sinclair's deranged state has exorcized the idea that 'haunted' him like a ghost. The sin of hubris and the desecration of Mary's body are avoided in the moment Musgrove recollects his own humanity, and he is saved after all. This would not have been possible, however, without the young physician.

The protagonist represents an ideal alternative to the character of the monstrous student. The narrative hints at least twice at the fact that the two received the same training. The young physician admits that Musgrove's plan did not shock him per se, as his training rubbed off the 'rust of a good many of [his] prejudices'.[44] The use of the word 'prejudice' may not be casual, since it had a key role in the anatomy debate. 'Vulgar prejudice' was the way in which supporters of the Anatomy Act defined the fear of dissection as a feeling entertained by the less educated portion of the population.[45] Even the better educated radicals, the ones who believed science was instrumental to social progress, perceived 'superstition or sentiment [as] emblematic of ignorance'.[46] The use of the word in 'The Dead Restored' suggests that the young physician himself is in favour of dissection. Moreover, when he voices again his disapproval of Musgrove's plans, Musgrove exclaims 'Oh, nonsense! [...] I never knew you were so scrupulous before',[47] which suggests that the young physician never before showed reservations concerning the dissection of a subject. Nevertheless, the young physician, differently from Musgrove, never behaves inhumanely. Although he sees beauty in Mary's body, he can control his morbid curiosity and put his respect for the grieving father and the dead herself before it. He retains, to an extent, the sensitivity and humanity students, through the study of anatomy, were supposed to lose, and he disapproves of the unnecessary, obsession-driven bodysnatching operation planned by his friend. His attitude towards Mary Sinclair's body suggests that he distinguishes between the legitimate pursuit of knowledge and transgression of social norms. For instance, Mary's body lacks the anonymity that, according to Rosner, helped medical students to achieve professional detachment.[48] The expressions the fictional medical student uses to refer to Mary, such as 'young creature',[49] indicate that to him Mary Sinclair is most decidedly not an anonymous subject. While Musgrove is entirely dominated by his hubris and morbid curiosity, the young physician can 'switch off' his clinical detachment depending on the situation. He exercises the decisional power that Foucault associated with the figure of the clinician to decide not to follow up to his medical gaze with the touch of the lancet. His gaze on Mary's body, unlike the necrophiliac gaze of Musgrove, takes the form of an aesthetic contemplation mitigated by his sympathy for Major Sinclair's sorrow and his respect for his daughter's remains.

Scruple and lack of scruple, therefore, is what distinguishes the human(e) medical student from the inhuman(e) one. This element is repeated in 'The Long Subject', in which the mature physician is juxtaposed to the figure of his friend, Dr. Garratt. Garratt, though less melodramatic a figure than Musgrove, embodies another element that characterized popular representations of surgeons, namely, cannibalism. He has a scientific interest in Long Hannibal's exceptionally tall corpse and he wants it as an 'interesting' specimen for his museum. The mature physician, while agreeing 'that the osteological system of [their] friend

Hannibal would answer the purpose very well',[50] voices his doubts on the feasibility of the plan. Garratt, however, is determined:

> Well, of course, there is some difficulty [...] We cannot expect that Mrs Jeffries would exactly like the skeleton of her husband to be hung up in a glass case in our new museum. People have their prejudices, you know, upon those matters. But still, when a great object is to be attained, I think we might take some trouble to carry it out. In a word, doctor, I have quite set my heart upon having Hannibal Jeffries nicely boiled, and scraped, and polished, and all the articulations of the joints got up with springs in the best manner, and hung up in the museum [...][51]

Whereas Dr. Garratt is almost a caricature, and some of his statements, such as the comment on Mrs. Jeffries 'not exactly liking' the idea of her husband's skeleton on display, are obviously meant as comic relief moments, his figure would have reminded the reader of some suspicious facts connected to the history of anatomy and dissection. The emphasis on Hannibal's skeleton's unusual height would have reminded the Victorian reader of the case of Charles Byrne, the Irish Giant. Byrne was part of the freak show scene of 1780s Britain; conscious that his colossal body had attracted the attention of many an anatomist, he lived in terror of being dissected after death. He asked for his body to be buried at sea, but John Hunter managed to obtain the body by bribery, and his skeleton is now exposed at the Hunterian Museum.[52] The story of 'Long' Hannibal Jeffries, therefore, connected the fictional surgeon Garratt to a real surgeon whose actions demonstrated that the medical community was not above legerdemain when an 'interesting' subject was the aim. Garratt is adamant on this point: 'when a great object is to be attained', one must 'take some trouble to carry it out', by which he means that he should either personally snatch the body or bribe someone to do it for him. During the narrative, Garratt attempts both. At first, he approaches the gravedigger with the intent of bribing him, but he stumbles on an honest man: Garratt informs the physician that he has 'sounded that official personage, and found it most emphatically and decidedly no go [...] [W]hen [he] began upon the subject of resurrection at all, [he] found such strong tide of prejudice against [him], that [he] saw it was in vain to battle with'.[53] It is possible to see that in 'The Long Subject' the word 'prejudice' is explicitly used as synonym of unreasonable, borderline-superstitious beliefs, a hamper on the progress of medical science.

Garratt's definition of the gravedigger's aversion as 'prejudice' identifies him as a medical man who considers the psychological attachment of the bereaved to the body and/or the reverence inspired by the dead, the mark of ignorance. Like John Hunter, he believes science should have no patience with, and no interest in, the wishes of the dead or their bereaved relatives.

Another element that connects Garratt to popular negative representations of the medical community is his detailed description of what he means to do with Long Hannibal's body. Boiling and scraping bones were operations linked to the butchery sector. Nineteenth-century surgery was a gruesome, messy, and violent business, profoundly different from nowadays focused, meticulous practice. The concept of anaesthesia was still in its infancy. The semi-conscious (when not fully conscious) patients were strapped down to the operating table, surrounded by men who held their limbs down for further measure, while the surgeon cut, explored, sawed or otherwise severed, and sewed back, perhaps lecturing medical students all the while. Consequently, surgeons and medical students were represented in cartoons as butchers, the instruments in their hands more similar to meat cleavers than lancets (Fig. 5).[54] Garratt's statement, then, assumes the macabre and disquieting undertones that characterized the perception of the surgeon anatomist as cannibalistic savages since the beginning of the century (as the allusive word 'feast' in the title of Rowlandson's print suggested).

To the comically gruesome Garratt is juxtaposed the figure of the mature physician. He appears to be more flexible about unauthorized exhumations than his younger self, insofar as they have a scientific purpose. When Garratt announces his intention of having Long Hannibal Jeffries' body 'nicely boiled, and scraped, and polished [...] and hung up in the museum' he replies: 'I have no objection', but then, he gives his own conditions: 'whatever means are taken to get the body, [...] the most special care shall be taken that Mrs Jeffries shall be kept in the most profound ignorance upon the subject'.[55] For the physician, the grieving widow's peace of mind is paramount. As in 'The Dead Restored', the physician objects to his personally disinterring the body, in this case on the ground of his acquaintance with the Jeffries. As in the previous narrative, his colleague tries to overcome his resistance by alluding to his familiarity with resurrectionism: 'Nonsense [...] You have been on these expeditions before [...] you will be invaluable'.[56] The physician relents; however, the mere fact that he voiced his scruples about exhuming

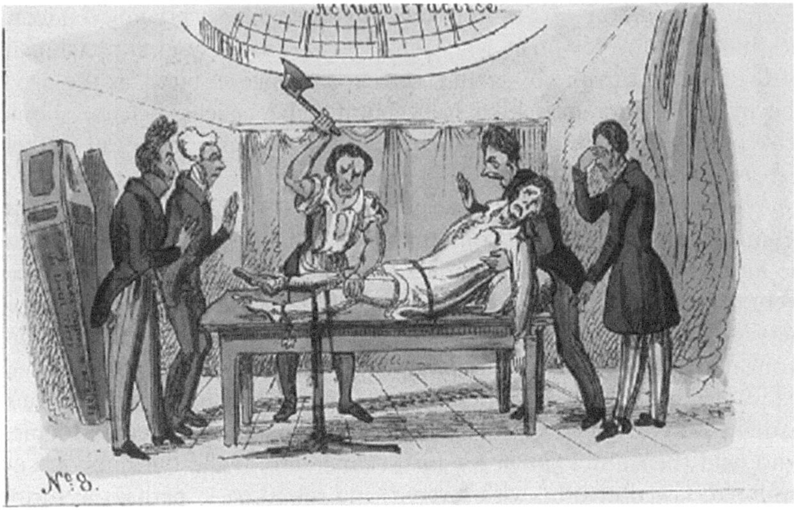

Fig. 5 Heath, William. Modern Medical Education: Actual Practice. Cartoon, 1825. By permission of University of Glasgow Library, Special Collections

a family friend separates him from the enthusiastic and cannibal istic Garratt. Furthermore, when Long Hannibal comes back to life in the cemetery, the physician is the only one who keeps his nerve and identifies the man as a patient coming out of a coma and not as a revenant. His past experiences as a resurrectionist disappear before the professional skill he demonstrates, as opposed to the comical loss of nerve Garratt and his exhumation party suffer in front of Hannibal's—literal—resurrection. Unlike them, the physician can use his medical gaze to 'see' the truth in what is happening: he is the only true medical man in the story.

There is a further aspect that makes the physician the medical man par excellence in the narrative. In this episode, and in others in the volumes, he shows exceptional stamina and prowess. The series recurrently portrays him overpowering his foes in fights or defending himself skilfully, and 'The Long Subject' is no exception. As the resurrectionist party is getting ready to set off for the cemetery, Garratt distributes the various tools, and the physician picks the spade 'because in addition to its usefulness, it is no bad weapon of defence'.[57] He is always ready to fight, although not in the position of the offender, which makes him appear both strong and gallant. His physical strength, as well as his agile intellect, is often devoted to the protection of the weak, as happens when he chases Josiah Gluck, the preacher, out of Mrs. Jeffries's house. The preacher, who introduces himself as 'one of the instruments of grace', has come to the house of the widow to 'convince [her] what a miserable wretch she is' and that 'her husband has gone to the bottomless pit'.[58] The physician dismisses Mr. Gluck's solemn declarations as 'unintelligible fanaticism' and kicks him out of the house, stating that 'if ever [he has] an aversion to anybody, it is when that body comes before [him] in the shape of a canting vagabond, who gets sleek and fat upon pretended sanctity, and goes about instilling fears into ignorant women and weak-minded persons'.[59] The character of the mature physician, therefore, doubles as science overpowering superstition: he is the scientific man who, enlightened by his studies, chases away manipulative bigots, preventing the more exposed members of society from becoming the hapless victims of fear. In brief, from 'The Long Subject', the mature physician emerges as a portentous combination of strength, stamina, mettle, and humanity. His perfection makes even his less amiable traits (such as his support of dissection) justifiable and forgivable. This extraordinary character represents the ideal doctor, a super-heroic doctor. He embodies the positive clinical gaze and clinical detachment, merging exceptional medical skill with the capacity to empathize and sympathize with the members of his community, to the point that he literally fights for their right to peace of mind.

Manuscripts reiterated fears that characterized the relationship between the public, particularly the lower class, and the medical fraternity. Granted, the themes of madness and monstrosity, translated in disturbing images of necrophilia and cannibalism, constituted excellent material for the sensationalistic plots relished by the audience of this genre. However, a close examination of the narratives shows active engagement with the anxieties sparked by the development of modern medical concepts such as the medical gaze, clinical detachment, and the figure of the clinician. It is also possible to identify elements from artistic representations of these medical concepts, particularly from the popular artistic form of the cartoon. Therefore, reading the figure of the physician merely as a sensational element would be reductive. Contrary to the general assumption that penny bloods were solely a commercial enterprise propelled by the newly developed serialized publications industry, this genre participated in debates concerning their readership: it adapted its stock characters—the monstrous villain and the brave hero, for instance—to the figures and anxieties concerned in the debate, in this case, the relationship between medicine and the poor.

I would take this point a step further and argue that *Manuscripts* proposes its own solution to the debate, a narrative alternative to the idea of the voyeuristic, cannibalistic anatomist that dominated popular representations. The two narratives do not engage with the fears the medical man awakened in the pauper simply by catering to the sensational potential of the monstrous medical student/man, as would be expected if financial return and cheap titillation really were the only propeller of the series. *Manuscripts*, instead, builds the character of the heroic doctor, the ideal medical man, a balanced student who becomes a humane practitioner, a figure in which stamina and strength blend with ethics, creating a fully trustworthy surgeon, entirely dedicated to the wellbeing of the community in which he operates, including—and principally—the poor. In this way, I argue, *Manuscripts* attempts to exorcize its readership's fears. The heroic doctor is the narrative antidote to the medical bogeymen that the lower class positioned at the centre of its subordinate situation as patients/subjects. This figure, modelled on the larger-than-life hero typical of penny blood narratives, conveys through the page the sense of safety regarding medical treatment that readers of penny bloods could not find in real life.

Once these considerations are made, it is possible to turn the analysis towards the discourse underlying the two narratives. The characters of Musgrove, Garratt, and the young and mature physician embody different interpretations of the figure of the medical man, particularly of the idea of power connected to it. Indeed, as medical historians have amply demonstrated, the situation immediately preceding the Anatomy Act, as well as its aftermath, pivoted on power relations among social strata.

2 Powerful Doctors and the Ethics of Exhumation: Solving a Deontological Issue

The dialectic of monstrosity and humanity in the episodes examined rests mostly on the perception the medical student/man has of the power that comes with his position as a member of the medical fraternity. The kind of power a nineteenth-century surgeon or physician could wield was compatible with the idea of knowledge-based power theorized by Foucault. While few members of the medical fraternity were rich or belonged to nobility, the knowledge of human anatomy and the ability to heal its ailments put every medical man in a position of power with respect to the other members of his community. Foucault underlines the role of pathological anatomy as the defining essence of the new medical expression that developed in the nineteenth century[60]; as can be evinced from a survey of early-nineteenth-century medical figures, the rise to prominence of pathological anatomy as a subject began early in the century. Thomas Southwood Smith, a supporter of the Anatomy Act and author of the pamphlet 'The Use of the Dead to the Living', pointed out that '[d]iagnosis of diseases of the epigastric region' had significantly advanced as dissection increased the knowledge of this area of the human body.[61] Dissection also allowed Robert Knox, the surgeon who purchased the bodies from the Edinburgh burkers, to prevent a patient in the Edinburgh Infirmary from undergoing potentially deadly surgery.[62] However, dissection's rise to prominence posed deontological problems that characterized the discussion of anatomy training for the best part of the century.

The use of paupers' bodies to acquire new medical knowledge was the core point of the deontological discussion. The power relationship between the medical fraternity and the poor was extremely unbalanced. Doctors occupied a higher position than the poor in the social system; to the social disparity corresponded an economic disparity, since the poor depended on charity treatment from the medical fraternity. This power disequilibrium allowed the medical fraternity to exploit the economic dependence and social inferiority of the poor, turning their bodies into teaching material. This was a hard bargain indeed for the poor, who suffered the consequences of receiving treatment from unskilled practitioners. A famous example is the case of the lithotomy Bransby Cooper, nephew of the more famous Sir Astley Cooper,[63] performed on Stephen Pollard, a labouring man, at Guy's Hospital in 1828.[64] While lithotomy, or bladder stone removal, was already a relatively simple operation, it proved an ordeal for Cooper, and resulted in the death of his patient the following day.[65] The *Lancet* covered the case as an example of singular medical incompetence, and blamed it on the nepotistic management of the London hospitals, the connection between Cooper and his uncle Sir Astley being, according to the magazine, the only possible reason why the incapable Cooper held a position at Guy's. Needless to say, Cooper sued Thomas Wakley (the *Lancet*'s editor) for libel. Richardson notes that during the trial, Sir Astley was called as a witness, and he candidly asserted that, although his nephew held a post at Guy's, he 'was neither fully skilled nor experienced'; he had no doubt, however, that in time Bransby would become 'a most thriving surgeon [...]'.[66] In brief, Sir Astley all but stated that the surgery Bransby operated on Pollard—and Pollard's death—were but part of his nephew's training.

Concerns about the possibility that poor patients may suffer from being used as training material for inexperienced surgeons were voiced also within the medical fraternity. Edinburgh surgeon Robert Liston remarked on the responsibility that came with treating poor patients for charity. '[W]ant of adroitness' with the lancet, and the 'unnecessary pain' and danger that came with it was in general deplorable, but it became 'highly criminal' when it affected patients in public practice, who had no choice but rely on charity treatment.[67] Liston, and his London colleague Charles Averill, argued that the only way to prevent this was to allow the medical students to practice at length on dead subjects. Of course, this raised yet more ethical issues, as the bodies that the medical community proposed to use were those of the very poor about whose safety they expressed concerns. This idea, which constituted the foundations of the Anatomy Act, ultimately enforced the power relationship already in place. Either by exploiting the pauper's socio-economically inferior position, or by compelling them to give up their remains with a law, the medical fraternity turned paupers' bodies into teaching material for its students.

The two episodes from *Manuscripts* here examined address directly the deontological issues of the pauper–medical student relationship. The narratives show medical men reacting in opposite ways to the idea of dissecting a pauper's body. In both cases, the exhumation and dissection represent an abuse of power, aggravated by the fact that they are not aimed to instruct a surgeon, but to satisfy his whims. The two narratives appear to be balancing the abuse of power inherent to the use of the body of the pauper for dissection against the importance of the role of the medical fraternity in society. While they cast a scrutinizing look on how surgeons use (and abuse) the power with which their knowledge invests them, they also display awareness of the necessity for well-trained medical men.

In 'The Dead Restored', Musgrove manifestly conceives the power of the medical man as based on social class. He claims that his 'profession'—although, in fact, he is still a student—grants him access to the house of Major Sinclair, who is a pauper, and therefore occupies a lower station in society. Musgrove's idea that his professional status constitutes a 'passport' does in fact indicate that he identifies himself with a socially powerful category, one whose members traditionally feel entitled to access the pauper's body. The fictional medical student is aware of his superior socio-economic position. Hence, he feels entitled to access the pauper's house and, by extension, to 'possess himself' of the corpse of his daughter—regardless of her being, technically speaking, a 'claimed' body—as if both were his own property. Musgrove, then, replicates on the page the power relationship between the medical trainees and the poor. While the series acknowledges the medical man's socio-economic and an intellectual superiority (after all, the protagonist is a physician, and in 'The Dead Restored' both students appear better educated and richer than Major Sinclair), the character of Musgrove criticizes the conception of medical knowledge as a source of social power. He is a medical student, a trainee, driven mad by the potential power—healing power, but also social and economic—that the achievement of further knowledge would give him. In this way, 'The Dead Restored' conflates Victorian concerns about the impact of dissection on the mental health of medical students with the greater debate over the deontological implications of the power the medical fraternity, down to its trainees, exercised on the poor.

The young physician, by contrast, embodies an entirely different interpretation of the power that comes from medical knowledge, because he embraces values other than the socio-economic power characterizing the medical fraternity, which makes him the ideal medical trainee. His knowledge of the human being appears to be far superior to Musgrove's because it transcends anatomy. He can sympathize with Major Sinclair's distress, and at the same time he invites him to 'be more a man',[68] that is, not to let himself be overcome by his feelings. Medical studies have strengthened the mental fibre of the young physician, although not to the point of dehumanizing him, which makes him more powerful than the arrogant, mad Musgrove. He restores Mary Sinclair to health by exercising both his medical knowledge and the authority with which this endows him. As he takes care of the child, he also takes charge, dispatching Musgrove to alert the village and instructing the first woman he meets to 'give [Mary] some weak wine and water, and a warm bath'.[69] The young physician's humane approach makes him a support to his community. He knows how to care for both the living and the dead (or nearly dead), and he knows how to heal both the mind and the body, to the point that he can reconnect his fanatical colleague to his feelings by showing him the grieving Major Sinclair weeping in the churchyard. He stands for the medical man who embraces the social responsibilities that come with the power the position of surgeon gives. This explains why he is against Mary Sinclair's unnecessary exhumation.

The position of the young physician on Musgrove's plans for the corpse of Mary underpins the medical discourse of 'The Dead Restored'. At the beginning of the narrative, it appears obvious that the young physician has been educated to consider exhumations necessary. Yet, he draws a line between acceptable and unacceptable exhumation, where unacceptable means it is based on the mere satisfaction of voyeuristic curiosity, and therefore represents an abuse of power. After all, voyeurism, that is the pleasure that comes from satisfying illicit curiosity—in this case towards the dead—is something that belonged to the purveyor of the series, which, as all penny bloods, catered to the reader's penchant for the excitement of 'seeing' the blood and the dead. The physician's curiosity towards the dead, instead, is reassuringly policed by his moral code and is bound to his need for instruction. This is one of the chief points of distinction between *Manuscripts* and Samuel Warren's *Passages*.

In the episode 'Grave Doings' of the *Blackwood's* series, which addresses the issue of resurrectionism, the physician-narrator tells of his first bodysnatching experience in jocular tones, as if recalling a goliardic experience. In this case too, the subject is a pauper girl—an unnamed young woman who entered the hospital and was subjected to several (experimental?) treatments to no avail: her symptoms baffle the surgeons, who intend to dissect her as soon as she will die to satisfy their curiosity as to her illness. Intelligence of this leaks to her brothers, who firmly oppose the surgeons' plan and take their sister home. Before the brothers' outrage, Warren's physician's reply is surprisingly cocky: 'Well, if you are afraid only of our dissecting her, we can get hold of her, if we are so disposed, as easily if she die with you, as with us'.[70] This brazen attitude, which becomes outright bullying when he points a pistol at the head of the Irish porter who is having qualms during the bodysnatching operation, characterizes him throughout the narrative. Furthermore, the story consistently shows a medical fraternity that considers bodysnatching, not just as a necessary evil, but as a test of mettle among students and teachers,[71] and this portrait is not undermined by the narrator's intermittent recollection that what they are performing is an 'unhallowed' deed.[72] The operation is successful in the end, and the narrative ends with a note from the narrator where he announces, not without complacency, that while the girl's relatives still believe her to be in her coffin, her skeleton 'adorns' a doctor's surgery and 'a preparation of her *heart* enriches—'s museum!'[73] Warren candidly represents surgeons as a powerful category, for whom snatching a body to satisfy their professional curiosity is legitimate, even if it includes deception and disregard for the pauper's explicit will. *Manuscripts*, on the other hand, written with the working-class audience's experience and sensitivity in mind, seeks a recalibration of the power relationship between its readership and the medical community, in which the poor are not at the mercy of a fraternity of power-intoxicated butchers, but in the hands of knowledgeable healers who use their authority wisely. The use of the bodies of the pauper for dissection can be imagined to be *useful*, yielding medical men who are simultaneously humane and proficient.

The concept of 'useful' dissection, for medical instruction and medical advancement purposes, emerges even more decidedly in the figure of the mature physician in 'The Long Subject'. The narrative's background are the anatomy wars, that is the moment in which shortage of bodies 'compelled' the surgeons 'to get possession of the dead, in order that they might obtain *useful* knowledge for the living'.[74] This opening is very different from that of 'Grave Doings', in which Warren's doctor petulantly appeals to his reader asking: 'You expect us to cure you of disease, and yet deny us the only means of learning *how*? You would have us bring you the ore of skill and experience, yet forbid us to break the soil, or sink a shaft! Is this fair, *fair* reader? Is this reasonable?'.[75] This concludes the discussion of the ethical implications of bodysnatching in 'Grave Doings'. *Manuscripts*' physician, instead, takes great pain to convince its readers of the 'usefulness' of dissection. His opening explanation of the surgeons' resurrectionist activity during the anatomy wars show awareness of his readers' feelings towards bodysnatching and dissection, as he defines bodysnatching as an '[attack] upon the sanctity of the grave' performed by 'too ardent sons of Esculapius'.[76] This representation of the surgeons as somewhat excessively zealous in pursuing 'useful' knowledge implies the will to consider a working-class reader's viewpoint. The fictional doctor, as if aware of the tension between his profession's interest and those of Long Hannibal and his family (and the readers), approaches the topic with caution when he explains Mrs. Jeffries that it was resurrectionism, in fact, that saved her husband's life. He is so convincing that Mrs. Jeffries openly rejoices that the physician behaved like 'a resurrectionist – a bodysnatcher, and all that sort of thing'[77]; had he not, the narrative implies, Hannibal would have died a most horrible death in his coffin. The physician, without recurring to Warren's doctor's plaintive tones, conveys a similar message to his own readers: there cannot be medicine, healing, or good physicians without the study of anatomy. So, anatomy is *useful*, a not 'well understood' subject on which, nevertheless, 'all classes of society' can learn to acquire 'more liberal' views, thanks to the 'march of education'.[78]

It is impossible to miss the influence of Utilitarian discourses and vocabulary on 'The Long Subject', which makes it even more significant when examined within the context of the Anatomy Act debate. Richardson notes the heavy Benthamite influence on the Select Committee on Anatomy and, ultimately, on the passing of the Act[79]: Jeremy Bentham, one of the fathers of Utilitarian philosophy, actively promoted the anatomists' cause in Parliament. The knowledge produced by studying anatomy and the subsequent improvement of medical treatment tallied perfectly with the 'useful knowledge' principle conducive to the Utilitarian goal of achieving happiness, if not for all members of society, at least for the greatest proportion of it. This idea, when applied to the Anatomy Act, presents obvious fallacies. First and foremost, the pauper, whose objections to be dissected should have been disregarded because the greater good of improved medical education was the goal, would be the sole source of bodies for dissection; simultaneously, they would be prevented from benefiting from these results, as their limited means would always compel them to resort to charity treatment. The upper classes, who would benefit from the Act's results instead, constituted by far a smaller proportion of the population. This did not prevent Utilitarianists from conducting a relentless campaign in favour of the Act, and to gather many followers; one of these was Thomas Southwood Smith, surgeon.

Smith is probably best known as the author of the pamphlet 'The Use of the Dead to the Living', first published as an article in the *Westminster Review*, and then reprinted as a pamphlet in 1824, to inform the members of Parliament who were to deliberate on the reforms concerning anatomy studies. The text argued in favour of the same solution proposed by the Benthamites, i.e. using the bodies of the pauper to solve the shortage of anatomy material supply. During the hearings of the Select Committee for Anatomy, Smith again made his point, and gave advice about how medical men should approach the pauper regarding anatomy. He stated that while he believed that '[the medical men could not] pay too much deference to the feelings of the poor' he found their aversion to dissection was 'neither so strong nor so difficult to be removed': in his career, he repeatedly overcame a dead pauper's relatives' objection to dissection 'by reasoning with [them], and explaining to them the importance of such inspection'.[80] To him, it was crucial that they should be present during the dissection because, he claimed, this invariably 'convinced [them] of its usefulness and importance'.[81] Smith's attitude towards poverty and dissection reverberates in *Manuscripts*' physician's language and behaviour. Unlike in Warren's *Passages*, where the impatient physician claims that he used 'in vain' 'all his rhetoric' on the girl's brothers, and finally brazenly tells them that he will dissect her anyway, *Manuscripts*' physician attempts to carefully explain to his (working-class) reader the 'usefulness' of dissection.

The fact that a penny blood should echo Utilitarian ideas can appear surprising, as they clashed with the audience's position before the Anatomy Act. I would suggest that this feature can be explained looking at the educational history of James Malcolm Rymer. Literary historians believe that he attended the Mechanics' Institution, which could have exposed him to the philosophy of 'useful knowledge', also in relation to medical studies, since the Institution had a role in the late-1820s' anatomy debate. Founded by George Birkbeck in 1823, Richardson relates that, in 1827, the Chancery building hosted a series of lectures on human anatomy given by Birkbeck himself, using 'a real human corpse'.[82] With this bold initiative, Birkbeck wanted to demonstrate that even the 'rough' men who attended the lectures at the Mechanics' Institution could understand, and benefit from, a lecture on science, and observers did indeed notice the audience's attentive attitude.[83] Richardson concludes that Birkbeck's lectures attest to the existence of a consistent group, composed of working-class people and intellectuals, sufficiently open-minded to reject aversion to dissection and who embraced the idea of 'the power of reason and the pre-eminent value of education'.[84] The view of dissection, and, to an extent, of resurrectionism presented in the narrative provides grounds to suppose that Rymer subscribed to these ideas. His profession, civil engineer, placed him on a better economic, educational, and social position than a factory worker, and therefore he might have been open to the idea of 'useful knowledge'. On the other hand, his work allowed him to meet his own audience and therefore he was likely to be aware of their feelings of powerlessness towards the Anatomy Act. Rymer's own professed responsiveness to his audience's tastes (as we have seen in the introduction) would explain the 'soft approach', similar to the one Smith promoted, in discussing the issue of anatomy and bodysnatching in the narrative.

'The Long Subject' reflects the tension experienced by the intellectual fringes of the lower strata of society when faced with the need to balance the grey areas of anatomy and dissection, against its undeniable (to an individual who could afford the luxury of an education) usefulness. The mature physician, like Smith, believes that the feelings of the pauper must be respected, and acknowledges the distress bodysnatching causes to the pauper describing the surgeons' 'attacks' to the graves as overzealous; however, he places greater emphasis on the fact that resurrectionism was done out of necessity. He alleges that medical students took part in bodysnatching only rarely and more for love of adventure than anything else, and he blames all the responsibility of the violation on corrupted sextons and gravediggers. By contrast, not only is there no trace in the episode—and indeed in the whole series—of 'professional' bodysnatchers,[85] but the gravedigger in the episode is incorruptible, while the body theft is planned and carried out by medical men. It is worth noting that the exhumation itself, which should have represented the climax of the narrative, though it is a realistic scene is significantly underplayed. When the physician and Garratt's party reach the cemetery, the operation of snatching the body of Hannibal Jeffries from his grave is dispatched in a few quick lines: the party works with '[great] expedition', 'suddenly' reaching the coffin, which is just 'tolerably' cleared of the mould before they set up to open it.[86] The only reason why Long Hannibal's gargantuan frame is not hauled up from its resting place any sooner is because Garratt's party is interrupted by that of Dr. James, another physician who also wanted the body for a specimen. This comic incident further underlines the medical community's excess of zeal and recalls the Irish Giant case: everybody wants Hannibal's body, as all London anatomists hoped to get hold of Byrne's. As in 'The Dead Restored', the narrative focalizes on the doctors' attitude toward the pauper's corpse, portraying a medical community prone to 'fight' over a body, while the titular physician renews his hopes, between himself, that Mrs. Jeffries should never discover the theft.[87] His character solves the conflict between the medical community's 'need' for Long Hannibal Jeffries' body and his right to a peaceful rest, as well as Mrs. Jeffries' right to peace of mind.

The mature physician's concern about Hannibal's widow defines him as a medical man who, though convinced of the importance of dissection, acknowledges the deontological issues it poses and the impact they may have on the community. The popular narrative thus solves the ethical conflict, representing a character whose strong deontological code moderates his views about dissecting the pauper's body. This character, set against a pre-Anatomy Act background, allows the readers of the post-Anatomy Act popular narrative to imagine that the use of paupers' bodies for dissection has a legitimate purpose and is policed by ethical people, and that it ultimately saves lives. In fact, as in 'The Dead Restored', the physician brings the person buried alive back to full health, unlike the medical characters who dismiss their ethical responsibilities. Through the character of the mature physician, moved by both Utilitarian principles and a strong deontological code, the narrative solves the tension between the unbalanced power relation between the poor and the medical community's need to instruct new surgeons.

The Anatomy Act underwrote the unbalanced power relationship between the poor and the surgeons that marked the pre-Act era without solving the deontological problems it posed. Historians have demonstrated that the poor did perceive the unfairness of this situation, contrary to the common belief that they were incapable of such refined sentiments.[88] The two narratives here examined produce a counterintuitive representation of these power dynamics, as they cast a new perspective on them by adding to the equation the character of the physician. The blend of Utilitarian principles and strong ethical code that define him stirs the attention from the act of bodysnatching towards the positive outcome for the subject who had been buried alive. The theme of premature burial, ultimately aimed at provoking a *frisson* at the thought of what would have happened if nobody opened those graves, did certainly fit the commercial strategies of the penny blood genre. However, these narratives deploy the theme of resurrection/revivification also to further address the ethical conflict inherent in the use of paupers' bodies for anatomy.

3 Liminal Spaces: Superstition and Rationality in Cemeteries

The two narratives approach the theme of the resurrected body with a rather different spirit: dramatic and dark in 'The Dead Restored', comic and thrilling in 'The Long Subject'. However, in both stories, this theme belongs to spaces connected to medicine and resurrectionism. The narrative spaces refer to geospaces that would have called to the reader's mind events related to the world of anatomy and the body trade, and both plots achieve their climax in the space of the cemetery. The representation of this latter space pivots on a dialectic paradigm that juxtaposes rationality and superstition: the superstitious terror that affects characters guilty of bodysnatching, with comic or dramatic consequences, is counterbalanced with the physician's courage, rationality, and medical skill. This paradigm exploits the association of the cemetery with death and burial rituals: it tapped into superstitions belonging to the cultural heritage of both the characters and the reader, and into contemporary anxieties about premature burial, recalled by the presence of bodies breaking the boundary between death and life. This reversion of the life-to-death movement, where bodies rise from the world of the dead below the ground to the world of the living above it, also addresses the legitimacy of opening a grave, and the consequences that befall the violators reflect the narratives' position on the matter.

'The Dead Restored' sets the character of the young physician, then a medical student, 'early in the year 18- [...] at the university of Edinburgh'.[89] As I noted before, the narrative space of Edinburgh in 'The Dead Restored' bears some connections with the geospace of Edinburgh, which the author reinforces by giving a Scottish accent to his characters and peppering their speech with terms from the Scottish vernacular.[90] The Scottish setting brings once again attention upon matters of authorship and Rymer's Scottish descent. A moderate acquaintance with the city's landscape and language, perhaps through tales heard from parents and relatives or visits to Scotland, may explain a relatively accurate geospatial representation (for instance, the existence of a slum in the proximity of 'the road to Leith') in which the gaps are filled in with fictional names such as Caulknows. However, and more importantly, such a setting would serve another purpose for a penny blood author: Edinburgh was the sensational setting par excellence. The city's architecture, with its winding closes where inhabitants were all but buried alive, favoured the circulation of chilling ghost stories.[91] Furthermore, the reader would immediately recognize the setting as the quintessential space of bodysnatching and murder, a brand imprinted on it by the Burkers case of 1828–1829.[92] Burke and Hare, however, were not the sole responsible for the Edinburgh-resurrectionism connection: the medical fraternity's behaviour was accountable for reinforcing this association, through such events as the Lasswade cemetery episode.[93] In February 1829, shortly after the Burkers case scandal exploded, three

Edinburgh anatomists purchased subjects from a gang of resurrectionists without bothering to ascertain their provenance. They came from the cemetery of the nearby town of Lasswade, where the resurrectionists, due to a combination of disorganization and poor time management, had left the tombs in a ravaged state that had alerted the town. During the investigations that ensued, the anatomists misplaced, made disappear, or disfigured the bodies, while refusing to answer questions in order not to incriminate themselves.[94] Their arrogance highlighted the privileged position of the Edinburgh surgeons, who were used to being 'treated very gently by the judicial authorities' in case of investigations.[95] These cases made early-nineteenth-century Edinburgh an arena where issues of ethics and power related to medicine and bodysnatching were consistently discussed, and Musgrove's obsessive hubris is compatible with the medical community as it had emerged from the Burke and Hare homicides and their aftermath. The Edinburgh setting of 'The Dead Restored' was therefore planned to remind the readers of the associations between medicine and crime that the press and social commentators had shaped about ten years previously. As for 'The Long Subject', it is set in a generic 'South of England',[96] possibly near London, since the Jeffries move there after Hannibal's 'resurrection'. While this setting shows a less immediate connection with anatomy and bodysnatching, the narrative, as did 'The Dead Restored', triggered this connection in the reader's mind by recalling real events with its plot. 'Long' Hannibal's skeleton is wanted because it is a curiosity, an interesting specimen; London was the home of famous collectors of specimens such as John Hunter and Sir Astley Cooper, and the space where the events related to Charles Byrne, the Irish Giant, unfolded. The emphasis on Hannibal's skeleton would remind the Victorian reader of the Irish Giant case.

The spaces of the two narratives, therefore, connected to geospaces that bore relationship with events involving death, medicine, and ethics. Although they set their action at the opposite sides of the country, there is a specific space that the two narratives share, both as a physical place and as an element of the story; that is, the cemetery. In general, the cemetery space marks the threshold between life and death and the irreversible downwards movement from one to the other: in this space, death is perceived in all its finality. Yet, in the two narratives, uncanny activity pervades the cemetery space, blurring the boundary between life and death: the dead who are there interred are not actually dead and they move back from death to life. The presence of the doctor is crucial at the point at which the 'resurrection' of the dead body assumes a literal meaning, rather than the figurative one implied in the term 'resurrectionism', suggesting that resurrectionism may yield positive outcomes. The cemetery thus becomes a space in which to discuss ethical issues, where the presence of the physician, crucial to the survival of the 'resurrected' person, rights the wrongness of desecrating a tomb.

'The Long Subject', through the physician's voice, defines resurrectionist activity as 'attacks upon the sanctity of the grave',[97] and the young physician in 'The Dead Restored' openly disapproves of his friend's intentions. This negative judgement of unnecessary exhumation resonates in the tomb violators' inability to cope with the sight of the resurrected corpse, which raises superstitious terrors in their minds. Indeed, Mary and Hannibal appear as ghosts in the eyes of the guilty observer, provoking intense reactions of fear:

> 'My God! what's this?' suddenly cried Musgrove.
>
> I looked around me [...] and the blood curdled round my heart for a moment as I saw the major's child sitting up with the grave clothes huddled around her, and her eyes wide open and fixed upon us.[98]

Mary's 'resurrection', eerie and abrupt, exploits the terror inspired by the child ghost. The suddenness of the unseen movement—the reader left Mary's body dead on the ground, and the next moment she is sitting upright in the rain—and the emphasis on the 'wideness' and 'fixity' of the open eyes, creates the impression that the supernatural is at work. Indeed, this is the impression it makes on the Major: '"No, no, Mary," he gasped. A gush of blood came from his mouth and he fell dead upon the grave mound'.[99] Major Sinclair does not consider for a moment that this may be a miraculous recovery, or resuscitation. His reaction rather suggests he believes that his own child, whose sacred repose he has disturbed, has come back from the dead to haunt him. The girl's fixed gaze suggests accusation, following a tradition that dates back to the staring spectre of Banquo in Shakespeare's *Macbeth*. The sight kills the Major on the spot, in a rather spectacular way that suits the dark, gothic atmosphere of the episode. Hannibal has a similar effect upon his resurrectionists. The first to witness his 'resurrection' is the mature physician. He is alarmed indeed, at first, but his reaction does not compare with the truly spectacular fright of Garratt and the other doctors. When they answer the physician's calls and reach the spot where the body is, they find that

'Hannibal Jeffries had risen to his feet, and looked, in the garments of the grave, at least double his usual height [...] he stood just where he was, and moved not the right or to the left'.[100] The sight of the revenant has a spectacularly comic effect: 'they with one accord turned and fled with a precipitation as ludicrous as it was in some respects alarming. They tumbled into the grave some of them [sic], and scrambled out again – over tombstones they went as if they were mad [...]'.[101] The grave robes that had 'huddled around' Mary's body, emphasizing the fragility of her age, affect the appearance of Hannibal in the opposite sense, turning him into a daunting, overwhelming presence. In both scenes, however, the shroud emphasizes the reversion of the movement from death to life: the girl's body seems to bloom out of it, and the grave robes invest Hannibal's appearance with solemnity. Both bodies stand perfectly still, as if retaining the stillness of death and increasing the impression—in both the reader and the observers in the narratives—that they are in the presence of a revenant.

The physical presence of the standing and staring bodies of Mary and Hannibal in the space of the cemetery triggers superstitious fears in the violators of their tombs. Their awareness of having done something *wrong*—something mad, in the case of Major Sinclair, and a selfish, anti-social act, in Garratt's case—raises in them the terror of the dead's revenge. Unlike the physician both young and mature, they have not 'rubbed off the rust of prejudice', which in the two narratives has a flexible meaning spanning from 'superstition' to its literal sense of 'being prejudiced', but that in general, I would conclude stands for 'ignorance'. Garratt too, notwithstanding his glorious plans for the skeleton, is put off by Hannibal's suddenly lively appearance. In 'The Long Subject' specifically, all the doctors on the scene show that they apparently did not benefit from the education they received and, confronted with something inexplicable, cannot cope with their cultural heritage relating to death and the dead. The space of the cemetery becomes, in the presence of the revenant dead, a space of judgement and indeed, they all receive punishment of the most terrible sort: they fill the graves they have emptied. In 'The Dead Restored', this happens literally: '[p]oor Major Sinclair was buried in the same grave from which he had taken his child',[102] as if to compensate for his sacrilegious act. A similar fate befalls the resurrectionist party of 'The Long Subject', although the scene is ludicrous rather than tragic. In their rush to flee from what they believe to be Hannibal's avenging spirit, they stumble and fall into his grave. Unlike Major Sinclair, their occupation of the grave is not final, as they 'scramble out' of it, but their fall, and their subsequent stumbling over the tombstones functions as a grimly comic memento mori. Anatomists were not indifferent to the idea of resurrectionism. Sir Astley Cooper, for instance, who died in 1841, disposed to be buried in a huge, impenetrable sarcophagus, which Richardson suggests indicates that he wanted to make sure not to fall into the hands of resurrectionists himself.[103] Therefore, the occupation, though brief, of the same grave of the corpse the surgeons meant to dissect works as a threat of retaliation from the supposedly vengeful revenant.

The two narratives punish the material agents of the inversion between the world of death and the world of life either with tragic death, or with the threat of death, which ludicrously exposes their lack of rationality and credibility. By contrast, the bodies of Mary and Hannibal, the actual displaced objects that undergo the inversion of movement, display an uncanny vitality. The uncanny vitality of burial grounds was a critical issue for most of the nineteenth century, as it was before, due to the unreliability of the methods to detect death. Placing a piece of glass in front of the mouth,[104] or feeling the pulse, were common methods.[105] This, Twitchell explains, caused people to be 'buried in comas, in catatonic fits, and in shock, especially during plague years, when the hasty disposal of the body was of primary importance'.[106] The cholera outbreaks of the nineteenth century led to a similar result. Cholera was particularly deceitful from this perspective, since, as Richardson points out, signs of life and death mixed in the diseased person: the muscular rigidity that affected the patient relaxed only after death, 'causing sudden convulsions that could be mistaken for signs of life'.[107] On the other hand, the blue colour of the skin typical of cholera patients, the stiffness, low temperature of the body, faint heartbeat and extremely slow breathing explained why sometimes 'people [...] survived a medical diagnosis of death'.[108] Examining the *Times*, articles covering news of people buried alive, or nearly buried alive, and of cases of mass hysteria about premature burial appear since the late 1700s and until well into the century, especially in periods of cholera outbreaks and, in general, periods associated with fever, both in England and abroad.[109] Tales of premature burial therefore imbued the geospace of the cemetery with the idea of uncanny activity. Premature burial also constituted an insidious danger for another category of people, that is, the gravediggers. The necessity of digging deep pits, either to bury multiple coffins, or to prevent resurrectionists from stealing the bodies, exposed them to the risk of a collapse of the sides of the pit, which would literally bury them alive. Sometimes, people nearby rescued the men in time, while others were less fortunate.[110] Considering the coverage that the issue enjoyed in the press, which is symptomatic of a deep anxiety, it is unsurprising that popular fiction should exploit the melodramatic effect of the character 'buried alive'. Mary and Hannibal's vitality in the narrative space of the cemetery turns this space into an uncannily lively threshold that connects life and death in a subversive way. Mary, who descended into the grave and was in succession beautiful corpse, medical subject, horrible sight, and terrifying revenant, and was, in brief, the supreme element of death in 'The Dead Restored', displays prodigious vitality and fertility. The story concludes with her nursing her fourth child. Although she hardly speaks in the whole narrative and she mostly stares, either through the stillness of her semi-closed eyelids as a corpse or through her wide-open eyes as a revenant, she turns from being the symbol of death to that of quintessential vitality: the fertile woman, the mother. Her stillness and speechlessness, which place her, like Hannibal, in connection with the world of the dead, contrast with her four children, which come into the narrative almost as suddenly and prodigiously as their mother came back to life in the cemetery. As for Hannibal, he comes back from the world of the dead a reformed man, an excellent husband and father who brings prosperity to his family through the wonderful inheritance of the uncle from the West Indies. His reward is a respectable position in society. The vitality of Hannibal and Mary, though, and the consequent transformation of the cemetery into a space of (uncannily) new life is possible only, thanks to the presence of the heroic physician in that space.

Conforming to the dialectic nature of the narrative, whereas the presence of the maniac or cannibalistic surgeon in the cemetery provokes an uncanny reversion of the movement from life to death, the presence of the physician, with his trained clinical gaze mitigated by a keen sense of ethics, is functional to restoring the resurrected patient to the world of the living. Indeed, he is the only rational medical man within the boundaries of the two cemeteries. In both narratives, he recovers from his surprise quickly enough to act and save the 'patients'. He is, at first, shocked in beholding the dead come back to life, and he describes his feelings with great precision. The upright, staring body of Mary Sinclair causes 'his blood [to curdle] round [his] heart',[111] and his agitation is even greater when he beholds Long Hannibal moving his head: 'I felt a rush of blood come up to my head, and then go back to my heart, while a strange, tingling sensation pervaded my whole system. [...] I don't know how long I remained transfixed [...]'.[112] However, in both cases, he quickly regains his rationality. As Major Sinclair falls dead on the cemetery ground, he 'starts to his feet': 'reflection had come to my aid, and I fully comprehended everything now. The child had only been in a trance, and the cold shock of the rain upon her face had restored her'.[113] Likewise, as Long Hannibal shakes his legs and groans, the mature physician exclaims: 'By the great God of Heavens [...] he lives!'[114]; by the time the whole company of failed resurrectionists has hurried away from the scene, he has recovered from the shock and formulated a diagnosis:

> The first shock of seeing him [...] move, had passed away. [...] I never was very superstitious, and when that shock had gone from my mind, I came at once to the conclusion that Hannibal Jeffries had been buried in a state of catalepsy, which had so strongly resembled death as to deceive every one.[115]

The physician, then, is not indifferent to the sight of a revivified corpse and to the idea, reinforced through the space of the cemetery, that it might be a revenant. Again, he appears to be in touch with his human side, to be close to the reader in this perspective. However, the triumph of his rational side at the sight of the dead makes him stand out among the characters in the cemetery. The young physician asserts that his education freed him from 'prejudice', that is, from that aversion to dissection that comes with the superstitious fear of the dead; as a mature physician, he specifies that he 'never was very superstitious'. Both in youth and in mature age, the physician overcomes his instinctive fear before the revivified dead and applies his rationality and knowledge. Indeed, he formulates a diagnosis immediately: 'trance' in one case, and 'catalepsy' in the other, which match the causes of premature burial enumerated by Twitchell.[116] He then acts quickly, a characteristic that distinguishes him in the whole series. The physician's prowess and mental agility is always evident, and in these two cases, he sets about the restoration of the non-dead, who immediately switch from the state of corpses to that of patients. The presence of the heroic and active, but simultaneously rational and cool-minded, physician in the space where the veil between death and life thins transforms the cemetery from a space of death, superstition, and ghastly bodysnatching activity into the space of positive action, rationality, and healing. The revenant's uncanny vitality becomes the triumphant vitality of the dead restored, making the cemetery a space of rebirth that provides a second chance for the revivified character.

The space of the cemetery, therefore, becomes a scenery for the battle of science against superstition, of anatomy versus prejudice. The dialectic structure that characterizes the two narratives at different levels concretizes in the inversion of the life-to-death movement and the uncanny vitality of the resurrected body. The agents of the inverted, uncanny movement undergo a trial and punishment that ultimately leads them to fill the same graves they have emptied, while the presence of the rational, 'good' physician transforms the cemetery into a space where rationality and medicine triumph. The narratives suggest, in this way, that the progress of medical science, which is obtained also through resurrectionism, is fundamental to fight, and win, the battle against death.

Manuscript engaged with several literary and artistic popular tropes to discuss issues of ethics and medicine. In so doing, the series tapped into the debate that saw the audience of the penny blood genre, as members of the working class, as the object of a deontological debate related to the study of anatomy. Certainly, making any conclusive assertion about authorial intention on this point would be foolhardy, considering the scanty information we possess on Rymer. Still, there is an obvious connection between the characters and events in the two narratives and the socio-cultural context related to the anatomy debate that in the 1840s and 1850s was still at its most vivid.

I would therefore venture that the series uses the elements characterizing the penny blood genre, such as stock, flat characters, melodrama, and sensationalism to translate for the audience the debate around the disposal of their remains, and to offer a narrative antidote to their powerless role in said debate. The construction of the positive character of the heroic doctor, or super-doctor, provides a medical figure who is both skilful and trustworthy, able to exercise his powerful medical gaze and simultaneously capable of retaining his humanity. This figure is juxtaposed to medical figures that reiterated tropes of madness and monstrosity founded on the distrust of the recent development in medical studies. The narrative thus allowed the reader to imagine to be under the protection of a wise and kind healer. Even the perspective of dissection after death is no longer terrifying, when it is framed within the narrative of the capable and trustworthy physician, told from his own very pen. In the debate around anatomy and ethics, trust and truth interlaced deeply and played a significant role. Foucault argued that the clinical gaze endowed the medical man with a privileged relationship with truth; yet, in the popular mind, the figure of the surgeon was conspicuously untrustworthy.

The next chapter moves on to examine another penny blood from the 1840s that engages with issues of medicine and truth. *Varney the Vampyre*, unlike *Manuscripts*, exploits the theme of the supernatural by placing an actual monster at the centre of its plot. However, some episodes display a marked interest in the figure of the medical man, which develops through discourses of madness and heroism remarkably similar to the ones of *Manuscripts.* Furthermore, the figure of the vampire facilitates the discussion of the inversion of the movement from life to death at the core of the ethical issues raised by resurrectionism. After all, a vampire is, first and foremost, a resurrected cadaver.

Notes

1. The two volumes are dated 1844 and 1847.
2. Tabitha Sparks, *The Doctor in the Victorian Novel* (Farnham: Ashgate, 2009), 14–15.
3. Ibid., 14.
4. Imitation of successful narratives in the cheap serialized fiction world could sometimes consist in outright plagiarism. See Louis James, *Fiction for the Working Man 1830–1850—A Study of the Literature Produced for the Working Classes in Early Victorian Urban England* (London: Oxford University Press, 1963), 45–71.
5. Pamela K. Gilbert, 'Sensation Fiction and the Medical Context,' in *The Cambridge Companion to Sensation Fiction*, ed. Andrew Mangham (Cambridge: Cambridge University Press, 2013), 187.
6. Anne-Marie Beller, 'Sensation Fiction in the 1850s,' in *The Cambridge Companion to Sensation Fiction* (Cambridge: Cambridge University Press, 2013), 6.
7. Elizabeth T. Hurren, *Dying for Victorian Medicine—English Anatomy and Its Trade in the Dead Poor, c. 1834–1929* (London: Palgrave Macmillan, 2012), 81.
8. Ibid.
9. Sparks, *The Doctor in the Victorian Novel*, 13.
10. Rymer, *Varney the Vampyre; or: The Feast of Blood*, 15; and Elizabeth James and Helen R. Smith, *Penny Dreadfuls and Boy's Adventures—The*

Barry Ono Collection of Victorian Popular Literature in The British Library (London: The British Library Board, 1998), xiv–xvi.
11. Gilbert, 'Sensation Fiction and the Medical Context,' 185.
12. Rymer, *Manuscripts*, 1:4.
13. This point is further explored in the next sections of the chapter.
14. Hurren, *Dying for Victorian Medicine*, 93–104.
15. Ibid., 105.
16. Ibid., 93–104.
17. Brock's biography of Hunter notes his achievements as a physician, anatomist, and man-midwife. Helen Brock, 'Hunter, William (1718–1783),' *Oxford Dictionary of National Biography*, accessed December 3, 2015, http://www.oxforddnb.com/view/article/14234. He and his brother John Hunter, surgeon and anatomist, were key figures in the medical environment of the second part of the eighteenth century. See Jacob W. Gruber, 'Hunter, John (1728–1793),' *Oxford Dictionary of National Biography*, accessed December 3, 2015, http://www.oxforddnb.com/view/article/14220. Both brothers deeply influenced medical practice and instruction in the following century, including the field of anatomy studies. John Hunter's massive collection of anatomical specimens was acquired by the government after his death and entrusted to the Company of Surgeons (soon to become the Royal College of Surgeons) and is still open to the public today in the Hunterian Museum in Lincoln's Inn Fields.
18. William Hunter, *Two Introductory Lessons Delivered by Dr. William Hunter to His Last Course of Anatomical Lectures, at His Theatre in Windmill Street* (London: J. Johnson, 1784), 67. Hunter's emphasis.
19. Ruth Richardson, *Death, Dissection and the Destitute* (Chicago and London: The University of Chicago Press, 1987), 31.
20. Michel Foucault, *The Birth of the Clinic* (London: Routledge, 2003), 2.
21. Ibid., 109.
22. Gilbert, 'Sensation Fiction and the Medical Context,' 183.
23. Rymer, *Manuscripts*, 1:4.
24. Ibid.
25. Ibid.
26. The first public anatomical museum was opened in London by the Italian Antonio Sarti in March 1839, and the central piece of Sarti's exhibition was a Florentine Anatomical Venus. Sarti believed in making the public privy to the world of anatomy, in instructing workers and women (who were usually excluded from anatomy studies) in the shape and functioning of the body. In this way, he maintained, not only would he spread knowledge on the 'laws of health', but he would also combat atheism, showing the divine at work in the human body. A.W.

Bates, '"Indecent and Demoralising Representations": Public Anatomy Museums in Mid-Victorian England,' *Medical History* 52, no. 1 (2008): 8–9.

27. Ludmilla Jordanova, *Sexual Visions—Images of Gender in Science and Medicine Between the Eighteenth and Twentieth Centuries*, 5th ed. (Madison: University of Winsconsin Press, 1989), 45.
28. Ibid., 44–45. Indeed, the habit of placing the female body in a reclining pose that simultaneously suggests that sensuality and vulnerability can be found in paintings representing reclining 'Venuses' as far back as the 16th century. Artists aimed to portray, and determine, the ideal female beauty, hence the name 'Venus' in the titles. Giorgione started the tradition with his 'Sleeping Venus' (1508), and several artists continued it in the following centuries, most notably his pupil and friend Titian, in his 'Venus of Urbino' (1538). The effort at representing the ideal female beauty for the (male) gaze to behold that characterized these paintings survived in the Anatomical Venus.
29. Ibid., 45.
30. Jordanova notes that wax was the material of choice for the models because it reproduced both the texture and the natural colour of skin. Ibid., 45.
31. Ibid., 45.
32. Rymer, *Manuscripts*, 1:4.
33. Ibid.
34. Ibid.
35. Hurren, *Dying for Victorian Medicine*, 106–13.
36. Ibid., 109.
37. Jordanova, *Sexual Visions*, 98.
38. Foucault, *The Birth of the Clinic*, 135.
39. Judith Flanders, *The Invention of Murder—How the Victorians Revelled in Death and Detection and Created Modern Crime* (London: Harper Press, 2011), 41.
40. Ibid., n.p.
41. In Sarah Wise, *The Italian Boy—Murder and Grave-Robbing in 1830s London*, Pimlico 20 (London: Jonathan Cape, 2004), 179, Wise writes that the dandyish Richard Partridge, the anatomist who reported to the police the London Burkers Bishop and Williams, was known for peppering his lectures with inappropriate jokes. Also, the anatomist Joseph Carpue claimed personal acquaintance with some of the subjects he dissected. He once told his students that a certain skeleton was 'the prettiest girl [he had] ever known'. Ibid., 179–80.
42. Wise, *The Italian Boy*, 179.
43. Rymer, *Manuscripts*, 1:5.

44. Ibid., 1:4.
45. Richardson, *Death, Dissection and the Destitute*, 151.
46. Ibid., 152.
47. Rymer, *Manuscripts*, 1:5.
48. Lisa Rosner, *The Anatomy Murders* (Philadelphia: Universty of Pennsylvania Press, 2010), 155.
49. Rymer, *Manuscripts*, 1:4.
50. Ibid., 2:48.
51. Ibid.
52. Byrne's skeleton is still a matter of both study and dispute. The analysis of his DNA proved that his acromegaly, or gigantism, originated in a tumour to his pituitary gland. In December 2011, legal researchers started a campaign to consign Byrne's skeleton to the sea, as the Irishman himself wished. The campaigners pressed the Royal College of Surgeons to respect Byrne's wish, arguing that everything that could be inferred from the study of his remains had been discovered. The Director of the Hunterian Museum, on the other hand, argued that current studies on the tumour of the pituitary gland connect Byrne's skeleton 'to living and future communities'. Press Association, 'Royal College of Surgeons Rejects Call to Bury Skeleton of "Irish Giant",' *The Guardian*, 2011, http://www.theguardian.com/science/2011/dec/22/irish-giant-skeleton-museum-display. See also 'Skeleton of Charles Byrne the Irish Giant Should Be Buried at Sea,' *The Telegraph*, 2011, http://www.telegraph.co.uk/news/newstopics/howaboutthat/8971853/Skeleton-of-Charles-Byrne-the-Irish-Giant-should-be-buried-at-sea.html.
53. Rymer, *Manuscripts*, 2:52.
54. Note the clothing and the attitude of the surgeon, who resembles more a butcher than a doctor, the swooning patient and the terrified attitude of the other characters. Note also the two empty coffins on the left, ominously suggesting the outcome of the operation.
55. Rymer, *Manuscripts*, 2:48.
56. Ibid., 2:52–53.
57. Ibid., 2:55.
58. Ibid., 2:52. The evangelical preacher is a typical figure in works ascribed to Rymer. In *Sweeney Todd* we have Reverend Lupin and in *Varney* we have Mr. Fletcher. We may suppose the physician's opinion of the category may somehow mirror Rymer's own, as from the narratives emerges a decided dislike towards it, which however does not extend to Anglican clergy. The priest who officiates at the burial rites for Hannibal Jeffries is gentle, empathetic, and has the beadle remove Mr. Gluck, come to stir more havoc, from the burial ground. The three preachers, instead, are

shifty, hypocritical, and sometimes lascivious, which suggests that Rymer was a conservative as regarded religion.

59. Ibid.
60. Foucault, *The Birth of the Clinic*, 152.
61. Rosner, *The Anatomy Murders*, 151.
62. Ibid., 153. Thanks to his dissection of a subject with a tumour in her neck, Knox was able to correct the diagnosis of the patient from aneurism (an aneurism operation would have certainly resulted in death) into tumour.
63. Astley Cooper was created Baronet in 1821 after successfully performing an operation to the King's scalp to remove a cyst. He was one of the best surgeons of his times, an excellent operator and an innovator, who had several influential patients and covered important positions in the London medical fraternity. W.F. Bynum, 'Cooper, Sir Astley Paston, First Baronet (1768–1841),' *Oxford Dictionary of National Biography*, accessed November 9, 2015, http://www.oxforddnb.com/view/article/6211.
64. For a detailed account of the case, see Richardson, *Death, Dissection and the Destitute*, 44–51.
65. Cooper subjected his conscious patient to a long, painful operation, after which he started lecturing his students on the stone that he had—finally—managed to extract, heedless of Pollard, still tied to the operating table.
66. Ibid., 48.
67. Robert Liston, *Practical Surgery*, ed. George W. Norris (Philadelphia: Thomas Coperthwait, 1842), xi. Quoted in Rosner, *The Anatomy Murders*, 152.
68. Rymer, *Manuscripts*, 1:6.
69. Ibid.
70. Samuel Warren, *Passages from the Diary of a Late Physician* (Edinburgh: Blackwood, William, 1834), 322.
71. Upon informing their superior that they intend to steal the girl's body, he congratulates them on their 'spirit' in taking up the 'dangerous' challenge, and the narrator complacently adds: 'Was it not skilfully said?' before reflecting on the '*eclat* with which the successful issue of the affair would be attended among [their] fellow students'. Ibid., 323, Warren's emphasis.
72. Ibid., 329.
73. Ibid., 338.
74. Rymer, *Manuscripts*, 2:44.
75. Warren, *Passages*, 321. Emphasis in the original.
76. Rymer, *Manuscripts*, 2:44.

77. Ibid., 2:61.
78. Ibid., 2:44.
79. Richardson, *Death, Dissection and the Destitute*, 107–8.
80. 'Report from the Select Committee for Anatomy' (London, 1828), 86.
81. Ibid.
82. Richardson, *Death, Dissection and the Destitute*, 151.
83. Ibid. Richardson notes that the event attracted a mixed response from the government. Mr. Warburton MP applauded it as a step towards beating the 'prejudice', that is, the anti-anatomy feelings generated by ignorance and superstition, in the popular mind; Sir Robert Peel, on the other hand, regarded the matter with scepticism, arguing that none of the men who attended the lectures rushed to offer his body to science.
84. Ibid., 108. It is worth noting that the same categories constituted the radical faction that was trying to oppose the Act. The development of faith in the power of knowledge and in the scientific purpose of anatomy divided the radicals making their opposition to Warburton's Bills ineffective. See ibid., 152.
85. The word 'bodysnatcher' is mostly used in the series as an insult to a surgeon, usually the physician himself. One example is in the episode 'The Hope of the Family; or, a Slight Indisposition.' Rymer, *Manuscripts*, 1:157.
86. Ibid., 2:55.
87. Ibid., 2:56.
88. See Richardson, *Death, Dissection and the Destitute*, 150 on the supposed 'bestiality' of the poor respect to the other classes.
89. Rymer, *Manuscripts*, 1:3.
90. For instance, the old woman who is assisting Major Sinclair in his grief explains to the young physician and Musgrove that there is a subscription 'to place the poor bairn decently in the *kirkyard* […]'. '"kirk" n.,' *OED Online*, accessed September 14, 2015, http://www.oed.com/view/Entry/103634?rskey=n7UX4e&result=1&isAdvanced=false#eid.
91. Edinburgh folklore suggests that the idea of being buried alive conveyed by the peculiar architecture of the close took a quite literal meaning in the popular mind. Mary King's Close, one of the (allegedly) most haunted spots of the Old Town according to the section 'A Close Most Haunted?' in the website dedicated to it, *The Real Mary King's Close*, draws a considerable part of its legends from the belief that the inhabitants of the close had been walled up there during one of the plague outbreaks to prevent the spread of the disease. The historical veracity of this information is doubtful; nevertheless, the popular mind imagined the Close as a space where people could be entombed alive.

92. It is worth noting that Hare's lodgings, where some of the murders were perpetrated, was in Tanner's Close.
93. For a detailed account of the episode, see Rosner, *The Anatomy Murders*, 42–53.
94. Ibid., 48.
95. Ibid., 42.
96. Rymer, *Manuscripts*, 2:45.
97. Ibid., 2:44.
98. Ibid., 1:6.
99. Ibid.
100. Ibid., 2:56.
101. Ibid.
102. Ibid., 1:6.
103. Richardson, *Death, Dissection and the Destitute*, 117.
104. Ibid., 227.
105. These uncertain methods constituted a problem well before the 1800s. Indeed, premature burial was one of the elements that supported the belief in vampires: finding the occupant of a coffin in a twisted position, the shroud ripped and stained with blood upon opening a tomb was considered a symptom of vampirism. In fact, it was the result of premature burial and a testament to the last, desperate moments of the person. What is also interesting for our discussion is that historians argue that the absence of a body from the coffin, also considered another a sign of vampirism, was instead to be attributed to bodysnatching. Carol A. Senf, *The Vampire in 19th Century English Literature* (Bowling Green, OH: Bowling Green State University Popular Press, 1988); and James B. Twitchell, *The Living Dead: A Study of the Vampire in Romantic Literature* (Durham: Duke University Press, 1981), 18–19.
106. Ibid., 18–19.
107. Richardson, *Death, Dissection and the Destitute*, 227.
108. Ibid.
109. In 1809, in Blackfriars, London, a mob attacked a cemetery when an old woman claimed she heard noises coming from the fresh grave of a fever victim (see 'The Neighbourhood of Christ-Church, Blackfriars ...,' *The Times*, 1809, 2). In 1832, both in Birmingham ('Desperate and Fatal Affray,' *The Times*, 1832, 3) and Liverpool ('Cholera at Liverpool,' *The Times*, 1832, 3) cholera victims were disinterred upon suggestion that they might have been interred alive, while an article from 1825 reports a dramatic piece of news from a Munich journal regarding the premature interment of a comedian ('Buried Alive,' *The Times*, 1825, 3).
110. Two examples coming from the *Times* are the cases of the gravedigger of Acton Churchyard, Macclesfield, and Mr. Thompson, gravedigger

of St. Brid's churchyard in Fleet Street. The first man was extracted from the seventeen-feet-deep tomb by the mourners of the funeral that was about to take place ('Last Week an Accident Occurred in Acton Churchyard…,' *The Times*, 1830, 3). Mr. Thompson, instead, was crushed under the weight of the earth and also 'of a tier of coffins piled on each other nearest [sic] the churchyard wall' ('Fatal Occurrence in St. Bride's Churchyard—A Man Buried Alive,' *The Times*, 1841, 3).

111. Rymer, *Manuscripts*, 1:6.
112. Ibid., 2:56.
113. Ibid., 1:6.
114. Ibid., 2:56.
115. Ibid.
116. Twitchell, *The Living Dead: A Study of the Vampire in Romantic Literature*, 19.

List of Works Cited

Bates, A. W. '"Indecent and Demoralising Representations": Public Anatomy Museums in Mid-Victorian England.' *Medical History* 52, no. 1 (2008): 1–22. http://www.scopus.com/inward/record.url?eid=2-s2.0-47749146135&partnerID=tZOtx3y1.

Brock, Helen. 'Hunter, William (1718–1783).' *Oxford Dictionary of National Biography*. Accessed December 3, 2015. http://www.oxforddnb.com/view/article/14234.

'Buried Alive.' *The Times*. October 17, 1825, 3.

Bynum, W. F. 'Cooper, Sir Astley Paston, First Baronet (1768–1841).' *Oxford Dictionary of National Biography*. Accessed November 9, 2015. http://www.oxforddnb.com/view/article/6211.

'Cholera at Liverpool.' *The Times*. August 24, 1832, 3.

'Desperate and Fatal Affray.' *The Times*. August 21, 1832, 3.

'Fatal Occurrence in St. Bride's Churchyard—A Man Buried Alive.' *The Times*. March 05, 1841, 6.

Flanders, Judith. *The Invention of Murder—How the Victorians Revelled in Death and Detection and Created Modern Crime*. London: Harper Press, 2011.

Foucault, Michel. *The Birth of the Clinic*. London: Routledge, 2003.

Gilbert, Pamela K. 'Sensation Fiction and the Medical Context.' In *The Cambridge Companion to Sensation Fiction*, edited by Andrew Mangham, 182–95. Cambridge: Cambridge University Press, 2013.

Gruber, Jacob W. 'Hunter, John (1728–1793).' *Oxford Dictionary of National Biography*. Accessed December 3, 2015. http://www.oxforddnb.com/view/article/14220.

Hunter, William. *Two Introductory Lessons Delivered by Dr. William Hunter to His Last Course of Anatomical Lectures, at His Theatre in Windmill Street.* London: J. Johnson, 1784.

Hurren, Elizabeth T. *Dying for Victorian Medicine—English Anatomy and Its Trade in the Dead Poor, c. 1834–1929.* London: Palgrave Macmillan, 2012.

Jordanova, Ludmilla. *Sexual Visions—Images of Gender in Science and Medicine Between the Eighteenth and Twentieth Centuries.* 5th ed. Madison: University of Winsconsin Press, 1989.

'"kirk" n.' *OED Online.* Accessed September 14, 2015. http://www.oed.com/view/Entry/103634?rskey=n7UX4e&result=1&isAdvanced=false#eid.

'Last Week an Accident Occurred in Acton Churchyard …' *The Times.* February 09, 1830, 3.

Liston, Robert. *Practical Surgery.* Edited by George W. Norris. Philadelphia: Thomas Coperthwait, 1842.

Press Association. 'Royal College of Surgeons Rejects Call to Bury Skeleton of "Irish Giant".' *The Guardian,* 2011. http://www.theguardian.com/science/2011/dec/22/irish-giant-skeleton-museum-display.

'Report from the Select Committee for Anatomy.' London, 1828.

Richardson, Ruth. *Death, Dissection and the Destitute.* Chicago and London: The University of Chicago Press, 1987.

Rosner, Lisa. *The Anatomy Murders.* Philadelphia: Universty of Pennsylvania Press, 2010.

Rymer, James Malcolm. *Manuscripts from the Diary of a Physician 2 Vols.* London: Lloyd, Edward, n.d.

Senf, Carol A. *The Vampire in 19th Century English Literature.* Bowling Green, OH: Bowling Green State University Popular Press, 1988.

'The Neighbourhood of Christ-Church, Blackfriars …' *The Times.* July 28, 1809, 2.

Twitchell, James B. *The Living Dead: A Study of the Vampire in Romantic Literature.* Durham: Duke University Press, 1981.

'Skeleton of Charles Byrne the Irish Giant Should Be Buried at Sea.' *The Telegraph,* 2011. http://www.telegraph.co.uk/news/newstopics/howaboutthat/8971853/Skeleton-of-Charles-Byrne-the-Irish-Giant-should-be-buried-at-sea.html.

Warren, Samuel. *Passages from the Diary of a Late Physician.* Edinburgh: Blackwood, William, 1834.

Wise, Sarah. *The Italian Boy—Murder and Grave-Robbing in 1830s London.* Pimlico 20. London: Jonathan Cape, 2004.

CHAPTER 3

Coping with the Displaced Corpse: Medicine, Truth, and Masculinity in *Varney the Vampyre*

Extremely successful in mid-Victorian England, famous among both detractors and supporters in academia afterwards, *Varney the Vampyre; or: the Feast of Blood* (hereafter *Varney*) is the epitome of the penny blood. The swashbuckling and larger-than-life adventures of the villainous eponymous protagonist, the vampire baronet Sir Francis Varney, simultaneously show the love for sensationalism that would open the way to sensation fiction in the second half of the century and the deep Gothic roots of the genre. The plot is ridden with murky elements from the Gothic tradition, such as stormy nights, dungeons, found manuscripts, and hidden treasures; more importantly for the subject at hand, the figure of the mad scientist, the Frankenstein trope, is prominent in the first part of the series.

Helen R. Smith ascribes the authorship of *Varney* to Rymer, and the series was produced in roughly the same time span as *Manuscripts*, between 1845 and 1847. While medicine is not the central theme of *Varney*, the two chief sagas[1] of the series, the Bannerworth and the Crofton saga, give medical men a central role, presenting both positive and negative versions of the figure of the doctor. Both versions present the influence of other popular and literary representations of the medical man (such as the Frankenstein trope), as well as of mid-century discourse s about medicine. Of course, whereas in *Manuscripts* the physician provides scientific explanations to apparently supernatural resurrections, in *Varney*, the theme of science must come to terms with the

A. Gasperini, *Nineteenth Century Popular Fiction, Medicine and Anatomy*, Palgrave Studies in Literature, Science and Medicine,
https://doi.org/10.1007/978-3-030-10916-5_3

quintessentially supernatural presence of the vampire. I would argue, though, that the supernatural element the vampire embodies contributes to, instead of clashing with, the discussion of the 'rational' subject of medicine in the plot. The vampire is first and foremost a displaced corpse; therefore, its appearance next to a doctor is significant if we consider the mid-nineteenth-century debate around the relationship between doctors and resurrected corpses.

Dr. Chillingworth and Dr. North appear in the Bannerworth and the Crofton saga, respectively. The two sagas also open and close the series, and the narrative explicitly connects them, and the two doctors, at both thematic and structural levels. The first episode of *Varney* features a vampire viciously attacking beautiful Flora Bannerworth, in her own bedroom, during a stormy night. The Bannerworths, a noble family now living in reduced circumstances, is composed of Flora, her brothers Henry and George, and their mother. Suddenly, they find themselves prostrated by the repeated attacks of the vampire. Simultaneously, a Sir Francis Varney takes residence in the neighbourhood and seems bent upon purchasing Bannerworth Hall from Henry. Sir Francis bears an uncanny resemblance to a portrait hanging in Flora's bedroom, which represents an ancestor of the Bannerworths and the girl swears to be the likeness of the vampire who attacked her. This is, of course, the truth, although Varney denies it at first. He is hunting for the hidden treasure of Marmaduke Bannerworth,[2] who was Varney's accomplice in his highwayman days. Marmaduke escaped with the treasure they had hoarded together, leaving Varney to the executioner's rope. Varney relentlessly attacks Flora to drive her family out of Bannerworth Hall and search for the treasure undisturbed. Through the saga, he breaks into Bannerworth Hall, fights duels, and imprisons Charles Holland, Flora's fiancée, when he hinders his plans. The climax of the saga corresponds to Dr. Chillingworth's confession that he is responsible for resuscitating the vampire. When he was studying medicine in London, the family physician galvanized the body of a hanged felon in the hope of being the first to successfully revivify a body. When Varney came back to life screaming 'Death, death, where is the treasure?',[3] the ambitious student could not imagine that he had just unleashed a vengeful vampire on his friends. As with Musgrove in *Manuscripts*, Chillingworth is obsessed with Varney's body and describes his obsession as 'monomania'.

The Crofton saga closes the vampire baronet's long, successful career. Varney constitutes an interesting exception to nineteenth-century vampires as he is the first suicidal undead. Unable to cope with his un-life, he attempts suicide by drowning; however, in *Varney*, vampires are revivified by moonbeams, and so it happens that the two sons of the influential Sir George Crofton find Varney's body while out fishing and bring it to the local bone-house. There, the moonbeams revivify Varney (giving a capital start to the sexton, who Varney catches in the act of stealing his own rings). The vampire, enraged at the frustration of his suicide attempt, seeks revenge. He gains admission in the Crofton family posing as Mr. Smith, grateful for the Crofton brothers' help. Then, he attacks the Crofton sisters, particularly Clara, whom he kills. Not only is Clara one of the very few people Varney kills in two years of weekly publication, but she is also the only vampire he spawns. Clara's resurrection, and her annihilation at the hands of an unruly mob, constitute the climax of the saga. When Clara dies after Varney's attack, and during the uncanny events that precede her interment, Dr. North, the family physician, supports the Croftons. As the men of the family, particularly the father, start showing signs of mental weakness and unmanly distress, Dr. North manages to restore the order. Cool, detached, and energetic, he keeps the family in balance.

1 Vampires, Mad Scientists, and Heroic Doctors: The Inhuman and Super-Human Paradigm

Any discussion about monstrosity in *Varney* must necessarily consider the presence of an actual vampire in the plot. Critics tend to evaluate the other characters' greater or lesser degree of mischievousness taking Varney as a reference. For instance, Herr asserts that '[a]s Rymer's chapters involving the Bannerworth saga progress, many of the non-Vampire characters become more monstrous than Varney himself'.[4]

While I agree that human characters sometimes outclass the vampire in wickedness, I would argue that to appreciate the nuances of this process, it is necessary to distinguish between 'monsters' and 'villains', taking Varney's condition as uncanny lively body as a reference point. Although there are several characters who gradually become villains, truly monstrous characters connect with the unnatural inversion of life and death that constitutes the core of *Varney* as a vampire narrative. While Mr. Marchdale, the treacherous family friend who plots against Charles Holland to dispose of a rival to Flora's heart, is certainly a 'villain', as Varney himself defines him,[5] to make a monster in *Varney* it takes, I would argue, quite the opposite intention to a murder, that is, bringing the dead back to life, perverting nature. In *Varney*, the concept itself of monster as a 'repulsively unnatural' creature showing inhuman cruelty[6] is indivisible from that of life–death inversion. In the series, two characters correspond to this concept: the vampire and the mad medical student, who share the essential feature of the definition of monstrosity above, that is, inhumanity. While Varney is, literally, not human, and therefore expected to be vicious, Dr. Chillingworth is even more disquieting because he does not display overtly monstrous features such as fangs and unnaturally long life: he actively chose, as a medical student, to act monstrously, and therefore he generated a monster. This binds him, and Varney, to a popular narrative that medical and literary historians have connected with the changes in the medical field throughout the nineteenth century, that is, *Frankenstein*.

Although he is no combination of body parts, the meaning of Varney's vampiric nature relates to the context of anatomy discourse. Unlike such supernatural manifestations as ghosts, his presence in the world of the living is physical. He is undeniably 'here', moving and occupying space; he interacts with humans around him, intruding in their environment and affecting their lives. As an element from the world below subversively appearing in the world above, Varney represents a threat to the living, both physically and psychologically. Physically, he hunts the living and sustains himself on their blood. Psychologically, Varney is a revenant, and therefore not only does he hunt the living, but he haunts them. He haunts Bannerworth Hall and Flora Bannerworth, seeking revenge against Marmaduke Bannerworth, and in the Crofton saga, he haunts the Croftons seeking revenge against the disturbance of his remains. In the Bannerworth saga, his presence also haunts Dr. Chillingworth, who recognizes Varney as the hanged felon he galvanized back to life several years back: 'I suspected it, do you know [...] His face *haunted* me [...] – awfully *haunted* me; and yet [...] I could not identify it'.[7] Unwittingly, but effectively nonetheless, Varney haunts the doctor with a different revenge spirit: he is Chillingworth's nemesis, his personal Frankenstein's Monster.

By the time *Varney* was being issued, the British reading public was familiar with such characters as Dr. Chillingworth. Chris Baldick examines a story titled 'The New Frankenstein' that appeared in *Fraser's Magazine* in 1838.[8] Its protagonist was a medical student whose attempt at building a Creature by putting together pieces of brain of different illustrious characters produces a monster with monumental speech difficulties. Baldick argues that this narrative indicated the 'assimilation' of the *Frankenstein* story in the culture and the 'experiment and revision' it was undergoing.[9] This, he holds, marked the rise in fiction of the 'mad scientist cliché', which almost invariably concretized in a medical trainee with delusions of grandeur and dedicated to galvanic experiments, whose secluded lifestyle resulted in an 'insensitive', when not downright 'misanthropic', rejection of social conventions.[10]

Varney subscribes to contemporary practices of appropriation of the Frankenstein narrative and elaborates it in the figure of Dr. Chillingworth. As Varney's attacks on the Bannerworths become bolder, Chillingworth must confess to the Bannerworths that he had a role in putting Varney on their path. He explains that, when he was a medical student in London, he became obsessed, like Shelley's Frankenstein, with the idea of revivifying a corpse through galvanism, a scientific method that attempts to reanimate dead tissue using electricity. Herr notes that, as the use of electricity in bringing the Creature to life is implied in the text, galvanism is the same technique used in *Frankenstein* to bring the Monster to life.[11] Further connections between the two narratives emerge in Varney's physical appearance and in his role with respect to the doctor responsible for his revivification. The vampire's body resembles that of the Creature in some respects: he is tall, indeed his 'full height' is 'immense'[12] and, when in human form, he is characterized by 'a strange distorted look [...] that has arisen from a spasmodic contraction of the muscles, in consequence of his having been hanged'.[13] The monstrous traits peculiar to the vampire, such as the protruding fangs, the eerie gaze, and the bloodless skin, are matched in Varney's body by a certain disproportion of features that reminds the reader of Frankenstein's Monster. Moreover, in both stories, the doctor's hubris produces a monster who brings death to the world of the living. The relationship between creator and Creature, however, differs in one point in the two narratives: while the Monster becomes vengeful and starts spreading chaos following Frankenstein's rejection, Varney is entirely disconnected from Chillingworth, and is born, or rather re-born, in an enraged and vengeful state. Spreading chaos is innate to him: being a vampire, an entity that does not belong to life and yet lives, he is a naturally disruptive presence.

The Monster purposefully hunts and haunts Frankenstein, while Varney's path only happens to cross Dr. Chillingworth's. Nonetheless, the effects of the meeting are powerful. Varney is a walking reminder of Chillingworth's folly, and compels him to reveal his shameful past. Since vampire resurrection in *Varney* is usually quite peaceful, connected to nature and to moonlight, it is meaningful that in the first part of the series Varney, the chief displaced body in the narrative, is introduced to the reader as the product of a disastrous medical experiment. The scientific nature of the revivification Chillingworth performs on Varney's corpse gives a distinctive medical-scientific nuance to the presence of the vampire. His impact on the lives of the Bannerworths is the direct consequence of a medical student's experiments on his corpse.

The series develops Varney rather counterintuitively, transforming him from villain into hero. It is likely that the audience found the swashbuckling, cunning vampire baronet far more interesting than the good but colourless Henry Bannerworth and Charles Holland. Consequently, Varney progressively evolves into a rogue with a code of honour, while, conversely, other characters become increasingly villainous. Varney's transformation becomes evident when he starts empathizing with his victims. Although initially he ruthlessly persecutes the Bannerworths, he then takes pity on them. He frees Charles Holland, whom he held captive in a dungeon, and shows contempt towards the treacherous Mr. Marchdale, who tries to kill Charles before Varney can free him. This moment marks Varney's passage from villain to heroic rogue, culminating in his dramatic display of grief and empathy after a mob kills Clara Crofton. In front of Clara's mangled and unburied body, he exclaims:

> [I]s this my work? Oh, horror! […] I thought I had […] completely crushed dove-eyed pity in my heart, but it is not so, and still sufficient of my once human feelings clings to me to make me grieve for thee, Clara Crofton, thou victim![14]

Varney's pity generates the narrative's core paradox, according to which deviant humans are worse than the monster itself. Notably, among the first characters who correspond to this paradox is a representative of the medical profession, whose members deemed inhumanity a necessary part of their training, a skill to develop together with their clinical gaze. The account of good Dr. Chillingworth's behaviour as a medical student is indeed 'worth a chill' and tallies with the highly sensational style that characterizes the whole series; simultaneously, it taps into issues related to the medical world, both fictional and real, with which the readership was likely to be familiar.

The revelation is made in episode 77, well into the Bannerworth saga. Varney has already freed Charles Holland, displaying what Henry defines 'a strange and wild kind of generosity'.[15] The men of the family and Dr. Chillingworth are now trying to discover the reason of Varney's obsession with Bannerworth Hall, which the reader can imagine to be Marmaduke's lost treasure. Meanwhile, though, an old acquaintance that connects the vampire and the doctor has made his appearance in the neighbourhood and is likewise in pursuit of the treasure. This is the former London's common hangman and, perhaps, it is his presence that compels the doctor to tell his story. Chillingworth begins by reminding his friends that he had started his studies very late in life, and therefore 'was extremely anxious to do the most [he] could in a very short space of time'.[16] Unlike 'the young men who affected to be studying in the same classes as [him]self', Chillingworth did not concern himself with 'what they considered life in London', but 'was indefatigable' in applying to his studies: 'there was nothing connected with them which [he] did not try to accomplish'.[17]

By distancing himself from his younger, lazy colleagues, and stressing his dedication to medical studies, Chillingworth tries to put himself under the best possible light. His remarks echo contemporary representations of medical students in the popular press, which described them as a parasitic and turbulent category of city denizens. Hurren examines a caustic article published on *The Penny Satirist* as an example of such representations.[18] Grainer, the author, uses the word 'infested' to describe the neighbourhoods around hospitals where medical students rented houses,[19] which immediately classifies them as parasites. He then proceeds to describe the medical student in its various environments, 'at home', 'abroad', and in the dissection room. 'At home' the student is a shabby, cigar-smoking idle, devoted more to drinking than studying. When he goes out, he turns into an exaggeratedly elegant 'dandy about town',[20] wearing expensive clothes of Italian and French fashion. Finally, in the dissection room, he becomes a prankster clad in weird garments, bent over the more disgusting part of his studies.

Hurren connects this kind of representation to public concerns about the commitment of medical students, who appeared to neglect their studies (and the payment of their fees) and favour smoking and drinking instead.[21]

Dr. Chillingworth's statements would call to the reader's mind the figure of the idle medical student, which is, however, eclipsed by the gradual appearance of the actual frightening figure of the doctor's tale: Chillingworth himself. He emphatically declares: 'there was nothing connected with [my studies] which I did not try to accomplish'.[22] This statement, which should vouch for Chillingworth's zeal, denotes not so much enthusiasm as compulsive tendencies, which concretized in the attempt to accomplish the forbidden scientific endeavour, that of bringing the dead back to life. As in *Manuscripts*, *Varney* introduces the figure that embodied one of the wider public's darkest fears regarding medicine and anatomy: the obsessed, mentally unstable doctor-in-training.

Chillingworth explains that he had become acquainted with the work of a Frenchman who managed, through galvanism, to resuscitate a dead man who had then lived for five weeks. This intelligence 'inflamed [his] imagination' so much that nothing 'seemed to [him] so desirable as getting hold' of a fresh corpse to attempt a resuscitation.[23] He therefore approaches the public hangman, knowing that he could provide him with 'the body of some condemned and executed man, upon whom [he] could try [his] skill'.[24] His obsession with the idea of accomplishing a resurrection connects Chillingworth with Shelley's Victor Frankenstein: like him, Chillingworth plans to defeat death, and to rise to the level of God. Rymer's doctor shares Shelley's doctor's hubris, the sin of arrogance committed against the gods by the man who fancies himself their equal. Gilbert notes that the 'ability to transcend individuality' of mid-century fictional doctors (particularly mad-doctors) may lead them to claim for themselves a 'god-like right to judgement'.[25] Dr. Chillingworth's embodiment of the arrogant doctor tainted with hubris, I argue, pre-dates the surgeons of the sensation novel. The Gothic roots of the penny blood in general, and of *Varney* in particular, allow the narrative to subscribe to ideas surrounding the medical figure that were already in the air, and that would be fully explored and exploited in the sensation novel. Free from the aesthetic and moral legacy that characterized the form of the novel,[26] the lowly form of the penny blood was ready to elaborate the doctor of the Frankenstein narrative into an even darker figure, charged with the negative connotations the medical fraternity was developing in the popular mind as the century progressed.

Chillingworth, unsurprisingly, is a supporter of dissection, and at the beginning of his tale, he frames his scheme to obtain a body for his experiment within the context of early-nineteenth-century body shortage. His speech includes typical Utilitarian concepts underpinning the ideologies that espoused the Anatomy Act. He underscores the difficulties he experienced in procuring subjects for dissection when he was a student, asserting that 'all sorts of schemes had to be put into requisition to accomplish so desirable and, indeed, absolutely necessary a purpose'.[27] His definition of dissection as 'desirable' and 'necessary' recalls Sir Astley Cooper's statement before the Select Committee in the summer of 1828. When asked how the surgeons felt about the necessity of employing bodysnatchers to get dissection material, Sir Astley answered apologetically: 'we are obliged to employ very faulty agents to obtain a desirable end'.[28] Chillingworth's approach to anatomy connects him also to Dr. Robert Knox. Knox was not particular about the 'agents' who procured bodies for his lessons, as long as he could count on the steady supply of cadavers to boost his prestige as an anatomist. Reaching the 'desirable end' was paramount in his mind. In a way, therefore, Knox was as guilty of medical hubris as his literary counterparts, Frankenstein and Chillingworth, were. From this perspective, Dr. Chillingworth stands at the centre of a web of literary and historical characters who contributed to creating the discourse on medical studies, and he constitutes the narrative's input to such discourse. The detachment from both the dead and the living and the goal-oriented attitude that characterized the medical world surface in Dr. Chillingworth's perception of dissection as necessary and desirable, and in the means he adopted to obtain Varney's body. To achieve his purpose, Chillingworth employs his own faulty agent, that is, London's common hangman, a character almost as marginalized as the bodysnatchers were.

In his determination to achieve his personal 'desirable end', Chillingworth shows appalling disregard for the life of the felon sentenced to death. Varney was to be hanged for 'highway robbery'.[29] Hanging would break his neck, preventing reanimation, but Chillingworth obtains from the hangman the assurance that he would 'manage to let him down gently, so that he shall die of suffocation, instead of having his neck put out of joint'.[30] Herr observes that this detail highlights Chillingworth's 'questionable' ethics, as suffocation would have meant prolonged suffering, instead of a quick death.[31] His light-heartedness in adding further pain to someone's death implies that, to the medical student Chillingworth, the unknown felon is worth much more dead than alive. Only the experiments it will be possible to make on his corpse matter. Chillingworth's cruelty suggests that his character is a commentary on the ethics of medical experimentation. To mark the callousness of his behaviour, at this point of the tale, the student Chillingworth abandons the sphere of rationality. To the executioner's suggestion, he answers: "If you can but succeed in that,' said I, for I was quite in a state of monomania upon the subject, 'I shall be much indebted to you [...]".[32] From a distance in years, Chillingworth explicitly describes himself as 'monomaniac', that is, insane.

As in *Manuscripts*, insanity, or 'monomania', taps into issues of power, madness and necrophilia related to the figure of the medical student. The medical student Chillingworth's ambition is that of becoming more 'scientific' than men of science themselves. His aim is that of 'imitating the learned Frenchman, who had published such an elaborate treatise on the mode of restoring life [...] to those who were pronounced by unscientific persons to be dead'.[33] Besides being shorthand for laypeople, the word 'unscientific' in this context also suggests that Chillingworth deems those medical men not brave enough to challenge death to be as unscientific as laypeople. The medical student experiences a power delirium, comparably to Shelley's Frankenstein. He does not reflect on the outcomes of his experiment. As he is impatient to resuscitate his subject, the hangman asks him if he has any plans for the revivified man. Chillingworth's answer is as adamant as it is revealing: 'Not I'.[34] The hangman humbly suggests that he 'consider[s] that it is really worth thinking of', but Chillingworth can 'think of nothing but the success or the non-success, in a physiological point of view, of [his] plan for restoring the dead to life'.[35] He works 'with [...] a vigour that promised the most completely successful results, if success can at all be an ingredient in what sober judgement would doubtless have denominated a mad-headed and wild scheme'.[36]

Chillingworth's careless answer, 'Not I', and the use of the term 'physiological', bespeak his disconnection from the human dimension of the corpse. His comment suggests that he is uninterested in both the individual human being and the wider community, while he is entirely focused on the success of his experiment. Chillingworth's plans do not envisage a way in which the public will benefit from the experiment's results because not even the man he is trying to revivify interests him. The perception of the dead body as an object of medical experimentation defines the medical student Chillingworth and constitutes a pivotal element of his monstrosity. His character embodies what Emma Liggins defines as the ambiguity of the power wielded by the nineteenth-century medical community: power to heal, but also power to harm, a 'violent' power lacking ethical regulations.[37] Combining this hostile perception of the medical fraternity's liminal position with the popular figure of the mad medical student, *Varney* reiterates nineteenth-century anxieties about medical education and the values—or lack thereof—it underwrote.

The juxtaposition of Chillingworth and the hangman, a man empowered to give death who shows more consideration of life than the medical student who is trying to restore it, reinforces the portrait of a medical community unable to reconcile its practices with the values of the community in which it acts. From a Foucauldian perspective, the hangman is a social outcast, and therefore empowered to speak the truth. He sees the chief flaw in Chillingworth's plan, the lack of empathy between doctor and subject, and he points it out to Chillingworth himself, inviting him to consider the humanity of the corpse. When Chillingworth's repeated attempts at revivifying Varney fail, the hangman comments on his failures with a grim joke: 'I am afraid, sir, it is much easier to kill than to restore their patients with doctors'.[38] This sarcastic remark confirms

Fig. 1 Heath, William. Modern Medical Education: Practical Results—At Home. Cartoon, 1825. By permission of University of Glasgow Library, Special Collections

the conception of an ambiguous medical fraternity empowered to both heal and kill, and connects to the idea diffused among the poor, confirmed by mortality rates, that entering a hospital was more likely to result in death than in healing.[39] As explained in the previous chapter, charity treatment could become an opportunity for doctors to practise dangerous surgery or to test new cures on poor patients before practising them on private, that is, richer, patients.[40] The popular form of the cartoon offers proof of this perception. Richardson examines the cartoon 'Modern Medical Education: Practical Results – At Home', published in the *Glasgow Looking Glass* within the series titled 'Essay on Modern Medical Education' (Fig. 1). The print shows a hospital whose entire staff, from the 'fashionable' doctor to the nurses attending the patients, are skeletons.[41] The doctor in the foreground appears at ease in the depressing environment of the room, but Richardson notes that he also seems completely oblivious of the patient lying at his feet, partly fallen out of the bed.[42]

The connection between the medical world and death in the print, reiterated in the hangman's comment in *Varney*, drew force from the anxiety generated by the perception of the medical man 'at ease' with death as monstrous. Dr. Chillingworth embodies this image, and simultaneously makes it literal, as the ambiguity underpinning his character is made more grotesque by the inclusion of necrophilia in the portrait. As with Musgrove's determination to 'possess [him]self' of Mary Sinclair's body,[43] Chillingworth's speech when he discusses the chance of 'getting hold' of a corpse has a strong necrophiliac edge.[44] Necrophilia relates Chillingworth to the concerns about the inappropriate interest of student s in dead bodies illustrated in the previous chapter, but more specifically with the fear that students might conduct experiments over patients.[45] By describing his state as monomaniacal at the beginning of the tale, Dr. Chillingworth bluntly admits that conceiving and carrying out his project compromised his mental balance. The words he uses to describe his feelings about the idea of obtaining a corpse emphasize this compromised state. Human beings naturally tend to refrain from contact with dead bodies; the medical student Chillingworth, instead, not only is impatient of getting hold of one, but he becomes almost disturbingly poetic when he talks about it. As soon as he completes the setting of his scientific apparatus in the hangman's house, which, of course, is appropriately 'old' and 'ruinous looking',[46] he starts growing as impatient as a lover waiting for the visit of his sweetheart:

> '[A]t least another hour must elapse before there could be the least chance of my seeing him arrive, for whom I so anxiously longed. I can safely say so infatuated was I upon the subject, that no fond lover ever looked with more nervous anxiety for the arrival of the chosen object of his heart, than I did for that dead body.'[47]

As soon as he hears the cart of the hangman approaching, though, he '[runs] down the stairs to meet what ninety-nine men out of a hundred would have gone some distance to avoid the sight of, namely, a corpse livid and fresh from the gallows'.[48]

As his narrative progresses, Chillingworth's language gradually becomes explicitly necrophiliac. The expressions 'infatuated', 'fond lover', and 'chosen object of his heart' on the one side, and 'dead body' on the other contrast grotesquely. In his 'impatience' in running down the stairs, he resembles a lover running towards the 'chosen object of his heart', but then the doctor points out that most men would have been repelled by what he is so impatient to meet. Chillingworth emphasizes the wrongness of his desire, specifying that the object is 'a corpse livid and fresh from the gallows'. This lurid detail conformed to the penny bloods' sensational style and suited the audience's taste for the macabre. However, it also underscored the clash between sane human behaviour and that of the medical student, who not only does seek corpses, but is *attracted* by them. This detail addressed two concerns related to the world of medicine that Victorian fiction explored. Firstly, the issue of metal health. Discussing representations of male hysteria in sensation fiction, Gilbert notes that '[a]fter all, it was practically normal for a woman to be a little unbalanced, but the madman is a spectacle of horror as well as pity'.[49] To this, I would add that, in Chillingworth's case, madness is horrifying not only because it is a mark of unmanliness, but also because it impairs the healing power of the medical mind and increases its potential violence. Furthermore, Dr. Chillingworth's necrophiliac language reflects concerns related to the sexualization of the medical gaze over the corpse, which we have already encountered in *Manuscripts*. Commenting on Foucault's theorization of the progressive eroticization of the medical gaze on the patient across the nineteenth century, Liggins suggests that this process makes the 'social and medical dominance' of the medical practitioner similar to 'the sexual dominance of the murderer'.[50] She individuates a series of similarities in both actions—in the ability of both doctor and murderer to make the subject 'silent'—and in lexicon, as both figures are 'attentive, insistent, [and] penetrating'.[51] Although Liggins applies this definition specifically to the relationship between doctor and female subject,[52] I would argue that Chillingworth's speech suggests that the idea was not gender-specific. Possibly, the insurgence of 'feminine' madness in Chillingworth's mind provoked in him homoerotic feelings towards the male corpse on which he is about to experiment. Or, more simply, Varney's corpse stands for every corpse, silent and powerless under the necrophiliac gaze of the medical student as does the corpse of Thurtell in Rowlandson's print. The medical student is thus tainted with one of the greatest taboos of humankind, which singles him out as a different, monstrous Creature. It is certainly possible to connect this almost excessively grotesque character, with his blend of hubris and necrophilia, to the penny blood audience's taste for gore and the macabre; however, the emphasis placed on his necrophilia questions the genuineness of Chillingworth's claims on the 'necessity' of dissection. The narrative seems to suggest that the practice is rather a tool in the surgeons' hands, which allows them to indulge with the complacency of the authorities in their monstrous tendencies.

As the series builds up its (final) climax, medical men assume again a central role. Varney, under the assumed name of Mr. Smith, has gained admission into the Croftons' house and has attacked Clara. Dr. North, who is already in the house to attend to Clara's fiancé, who is recovering from a fall from his horse, takes control of the situation. It is discovered that North is a descendant of Dr. Chillingworth and, as his ancestor, his character shows clear Gothic influences: while Chillingworth is the re-elaboration of the Frankenstein narrative, North reveals his ancestry, as well as important information that could explain the Croftons' current predicaments, through the Gothic trope of the found manuscript. The doctor produces 'some printed papers' cut from 'a medical publication' containing an account written by 'a distant relation of [his], likewise a surgeon'[53]:

> 'The communication was so curious,' [North] said, 'that I cut it out of the old volume in which it appeared, and kept it ever since.'
>
> 'Pray,' said Mr. Smith, 'what was the name of your distant relation, the medical man?'
>
> 'Chillingworth.'
>
> 'Oh indeed; an odd name rather, I don't recollect ever hearing of it.'
>
> 'No, sir, it is not likely you should. Dr. Chillingworth has been dead many years [...]'[54]

In his work, Chillingworth illustrates the attack of the vampire named Varney on his family friends. North explains that Chillingworth 'of course suppressed names' in his story and declared that 'but for touching the feeling of living persons', he would have unveiled more 'curious particulars'.[55] The apparent similarities between this passage and *Manuscripts*' opening formula allow to further speculate on the influence of *Manuscripts* on *Varney* and its representation of medical men, most evident in the juxtaposition of the 'bad' and 'good' doctor.

Dr. North is an exceptionally cool, rational individual, whose remarkable character is ascribed to his medical training. When Clara Crofton is found dead, North immediately takes charge and, while he does display detachment towards the corpse, he seems to do so to maintain order in the family circle, as its men prove themselves unequal to the task. When he urges the girl's father and siblings to leave the room and the distracted father protests that he cannot leave his child, Dr. North answers categorically: 'That [...] is not your child'.[56] Suddenly, by the power of the doctor's words, who can speak truth and create truth, Clara Crofton ceases being Clara and becomes a corpse. Unlike Mary Sinclair in *Manuscripts*, or Varney's body in Chillingworth's story, her corpse does not fascinate Dr. North. In his view, she is now something separate, inanimate, an unfeeling object, which her family must not regard with affection. He sternly rebukes the distracted father: '[y]ou are in grief, sir, and know not what you say. These were not else the words that would fall from the lips of a man as you are'.[57] As with the young physician's exhortation to Major Sinclair to be 'more a man' before the death of his daughter, Dr. North takes upon himself the duty—and the power—to regulate the display of grief according to the Victorian norms that gendered it.

Excessive display of grief, as well as (and akin to) irrationality and lack of mental balance, was considered to be typically feminine, hence inappropriate in a man. Hurren relates the 1890 case of Mary Ann Huckle, a pauper who appeared in court for withholding the body of her dead husband, Tom, for four days and three nights, 'sleeping with [the] corpse'.[58] The judges finally put down her behaviour to 'eccentricity'.[59] Mary was thought, in brief, to be not entirely in her wits, which tallies with Gilbert's argument that mental unbalance was considered normal in a woman. The verdict in the Mary Huckle case was founded on the idea that, even in grief, one must display balance, of both demeanour and mind; her behaviour, consequently, must denote insanity. Examining the ideological biases and cultural stereotypes underpinning the gendering of disease in the nineteenth century, Wood underscores the medical community's efforts to frame male nervousness 'within constructs of masculinity'.[60] Doctors tended to fall back on gender stereotypes when defining nervous disease, in order to promote the new ideals of masculinity as characterized by vigour and self-control.[61] The result was that both medical and literary representation of male nervousness portrayed a patient 'femininized by the very nature of his disease'.[62] Wood adds that one important element that emerges from literary reiterations and challenges of this image of feminized nervous man is that, as the century progressed, medicine took it upon itself to establish what could be considered gender-appropriate features and attitudes.[63] I would argue that not only does Dr. North in *Varney* correspond to the idea of the doctor as responsible for establishing 'normal' masculine behaviour (as well as what constitutes deviation from such a norm) but also that his knowledge of the appropriate parameters of masculinity makes him the paradigm of appropriate manliness. His person, strong and tough-minded, is the physical embodiment of appropriate masculinity, which endows him with the authority to police the Crofton men's grief.

The Crofton men show a dangerously feminine amount of grief. Herr notices that 'as the serial progresses, the Crofton men's constitution becomes increasingly frail', while '[h]yper-sensitivity is generally linked with feminine characteristics', and this causes the family's destruction.[64] The feminization of the Crofton men tallies with the Victorian gender-determined behavioural codes, as well as with images of men emasculated by their own nervousness that appeared in Victorian fiction examined by Wood. To induce them to show some stamina, Dr. North explains to them that the appropriate (i.e. manly) way to grieve is to adopt a philosophical attitude and to show 'what amount of resignation you can' in front of 'that stroke of destiny which you cannot control'.[65] He notes that, while death might be 'an evil to [them] in [their] loss', is actually the end of all pain and suffering to 'her who has gone from [them]'; after all, he argues, 'she has but gone a few years [...] earlier than usual'.[66] The doctor's detached speech, in which he does not even call Clara by her name, has, surprisingly, an immediate relieving effect on the men of the family. The narrator declares that this kind of speech 'was sure to have its effects upon persons in the habit of conversing coolly and calmly upon general subjects', such as the Crofton men should be, and the doctor indeed succeeds in making them 'exhibi[t] the rational grief of men'.[67] He succeeds because he possesses the toughness that was thought to characterize the male mind. Unlike the neurotic medical student of the Bannerworth saga, whose mind is 'inflamed' by delusions of grandeur and necrophiliac thoughts, he is a true man showing stamina, whose mind is lucid and clear, conforming to the modern, un-romantic Victorian ideal.

Notably, the narrative seems to connect Dr. North's manliness to the clinical detachment he developed through his profession. The connection surfaces when he gets in physical contact with the dead, at which point his character assumes superhuman undertones. When Clara's corpse is found inexplicably lying on the floor, out of the bed where they had left it, the father and the two brothers are again frantic, helpless, at loss about how to react. Dr. North saves the situation once more, turning them all out of the room and then, 'having professionally lost all dread of the dead', he 'lift[s] the body upon the bed again, and dispose[s] of it properly'.[68] The 'professional' loss of the 'dread of the dead' is obviously the clinical detachment developed by surgeons through their anatomy training, and defines the moment in which North, physically 'lifting' Clara's corpse from the floor, appears simultaneously strong, brave, and sensible. Clinical detachment singles out the doctor from the other men in the Crofton saga: it frees him from the subjection to his nerves and makes him keep full control over his emotions before death, the supreme element of distress, which no other man in the narrative seems capable to do. Unlike the repulsive attraction that characterized Chillingworth's touch on Varney's body, North's physical contact with Clara's corpse is functional to showing the strength and coolness that the Victorians ascribed to the male, as opposed to the female, mind. I would venture that the narrative is suggesting that medical training is what made Dr. North a true manly man: his ability to exercise a clinical gaze over the dead body, unencumbered by nervous emotions, endows him with the 'power of decision and intervention' that Foucault identifies as the defining trait of the medical figure.[69] Consequently, the narrative puts him in charge of the situation, making him the leader of the family in the moment of crisis. The power mania displayed by the neurotic medical student Chillingworth is reversed into the positive physical and mental power the leader Dr. North embodies.

Notwithstanding the time lapse between the characters of Chillingworth and North in *Varney*, the narrative explicitly connects them making them distant relatives, calling to the reader's mind Chillingworth's past as a mad medical student just before showing the reader an entirely opposite kind of medical man. As with *Manuscripts*,

Varney presents a polarized view of the medical practitioner, which, among other things, would confirm Smith's conclusions on authorship. The similarity could be simply ascribed to the fact that penny blood plots tended to be repetitive. However, this would not explain satisfactorily the deliberateness with which the narrative connects, indeed juxtaposes, the two doctors. I would suggest that such deliberateness rather expresses an intention to engage with the mid-century debate around medical education.

Sparks notes that around mid-century, the medical fraternity was interrogating itself on the challenges posed by the developments in medical education. As surgery and the rise to prominence of anatomy made medicine increasingly invasive, the medical community found itself divided between upholding traditional medical education based on 'gentlemanly ideals' and endorsing a new type of education based on experimentation.[70] Terrie Romano underscores how the 'gentlemanliness' underpinning the classical education adopted in medical curricula thus far, which corresponded to the medical fraternity's aspirations to gentility, also defined the intrinsic manliness of the profession.[71] Both Sparks and Romano note that the debate reached the wider public, fostering anxieties about medical education's departure from the moral and ethical codes that regulated the rest of the society.[72] Simultaneously, as both Richardson[73] and Hurren[74] observe, the wider public was also openly expressing concerns about the possibility that surgeons might experiment with patients, especially poor ones. While the narrative form of the novel started engaging with these themes only later in the century, the penny blood genre's openness to disturbing and unsavoury themes allowed the debate on medical education to find its way into the pages of *Varney*. Chillingworth and North elaborate two opposite consequences of the 'necessary inhumanity', in Hunterian words, that characterized medical training. While Chillingworth, conforming to the Gothic cliché of the mad medical student, is unethical and obsessed, and consequently appears monstrous, North uses his clinical detachment to guide his community and reach his full masculine potential. He makes, thus, the jump from inhuman to superhuman, which singles him out from other men as a model to imitate and a leader to follow, rather than a monster to shun. The medical student Chillingworth—and, consequently, Chillingworth the doctor—cannot fulfil that potential, because irrational behaviour is perceived as an eminently feminine component. As a nervous individual, he can only spread chaos, and never re-establish order.

The hyperbolic nature of the penny blood genre re-elaborates the Frankenstein narrative into a monstrous medical student who grotesquely embodies Victorian concerns about madness and necrophilia, concretizing anxieties about deviation from ethics in medical practice and the rising field of experimental medicine. To his figure is juxtaposed the equally hyperbolic Dr. North, a super-doctor in the same fashion of *Manuscripts*' physician. Unlike in *Manuscripts*, in *Varney*, the contrast between medical men encompasses discourses of gender. While the mad medical student, unable to control his nerves, becomes a necrophiliac with distinct homoerotic traits, breaking in this way two taboos simultaneously, the seasoned practitioner is a paradigm of masculinity, which the narrative directly connects to his medical training. Consequently, he has the power to invigilate over the masculinity of other male characters, claiming for himself the role of leader.

I would suggest that this indicates that not only does the narrative reiterate the increasing authority of medicine in establishing the parameters of gender-appropriate behaviour Wood notes,[75] but it ratifies it, simultaneously juxtaposing it to a negative image of insane, unreliable, and unethical experimental science. This representation of the medical figure suggests that *Varney* took a position in the debate of gentlemanly versus experimental science discussed by Romano,[76] underwriting Victorian values of masculinity and gentlemanliness against the suspicious figure of the experimental scientist, whose power is perceived as dangerous because he is mentally unstable. At the same time, the narrative counterintuitively sublimates the 'necessary' inhumanity derived from the development of pathological anatomy as the origin of North's mental strength and of his success as an effective doctor and leader. Rather than espousing monstrosity, North transcends humanity, which empowers him to re-establish order where the undead brought chaos. The narrative thus attempts to balance the concerns about the ambiguous power of the medical fraternity to both hurt and heal, providing a narrative solution in the character of the superhuman doctor.

2 Truth and Ambiguity: The Language of Power and Medicine

In *Varney*, the ambiguity of medical power emerges also in the relationship its embodiments entertain with truth. The nineteenth-century doctor merged two important points that Foucault individuated in the relationship between society, knowledge, and discourse control, that is, the access to knowledge and the regulation of other people's access to that knowledge.

As explained before, the manifest increase of the power of the medical gaze to pierce into what was concealed, and the exclusivity of this ability, were a source of anxiety for the wider public.[77] In *Varney*, this aspect of the medical figure must be considered, as with monstrosity, against the presence of the supreme metaphor of power in the narrative, that is, Varney. A figure connected to abusive, tyrannical power in both lore and literature, the vampire manifests his ability to subjugate other characters by skilfully controlling the circulation of knowledge in the plot. Moreover, being undead, he embodies the uncanny reversion from death to life; it is worth noting that, in *Varney*, truth is at its most blurred when this reversion takes place, in which moment characters display greater or lesser ability to 'see' the truth. Within this frame, the narrative juxtaposes Chillingworth, whose madness impairs his clinical gaze, to the clear-sighted Dr. North. The Bannerworth saga also discusses the gaze on death and its ability to access truth in relation to a third element, that is, the uninitiated, ignorant mob, who, lacking any kind of medical training, cannot benefit from engaging visually with death.

Varney establishes himself as *the* liar in the story since the beginning of the Bannerworth saga. Every word Varney utters sounds dangerous, and he displays a singular ability in twisting words. The other characters attempt to defend themselves from him and outsmart him, always unsuccessfully. His first meeting with Henry Bannerworth, after Flora has been attacked, provides a clear example of the power disparity between him and the other characters. Meeting his new neighbour, Henry is aghast to discover that he is the living image of the portrait whose subject Flora swears to be her assailant. Henry is terrified, while Varney behaves with the greatest innocence and politeness; throughout the scene, the narrator stresses the baronet's 'bland, musical' voice, and the 'mellifluous' tone that 'seemed habitual with him'.[78] The greater the distress Henry displays, the greater the sweetness of Varney's voice. This is a red flag for the reader, and builds up a feeling of uncanniness, giving the impression that the vampire's voice anaesthetizes his listeners, frustrating and confusing them. A further eerie component emerges when Varney excuses himself from the partaking of the refreshments he offers his guests, explaining that he is 'under a strict regimen': '[t]he simplest diet alone does for [him], and [he has] accustomed [him]self to long abstinence'.[79]

While these statements have the ring of truth, the reader can catch their subtle second meaning (particularly in the comment on his diet, which is remarkably similar to the feeding habits of vampires, with which the English public was acquainted since John William Polidori's *The Vampyre*).[80] Consequently, the reader can connect the baronet to the vampire who attacked Flora. By contrast, the characters in the story cannot ascertain the truth and they must take Sir Francis' words at face value.

The vampire's innocent demeanour causes the other characters to doubt their own senses, as Henry does when Varney assures him that he has no connection with his family, although the resemblance between the vampire and the portrait is plain to the young man. Varney enjoys upsetting the other characters. He purposefully disregards Henry's earnest request never to visit Bannerworth Hall, because of his aspect, which may disturb Flora. When the girl meets him, she is terrified (as was arguably Varney's purpose), even though he addresses her with 'winning accents', 'bland tones', and 'courtly' manners.[81] After Flora repairs to her apartments, the Bannerworth brothers, Charles Holland, and Mr. Marchdale (who has not yet betrayed the family) try to beat him at his own game and make him expose himself by asking him tricky questions about vampires. Varney, however, is a much more skilled player and answers with an earnest gaze that, Charles suspects, suggests familiarity with 'cross-questioning'.[82] Meanwhile, the vampire taunts his victims, regretting Flora's departure and alluding to his love for 'young persons of health' and their 'rosy cheeks. Where the warm blood mantles in the superficial veins, and all is loveliness and life'.[83] When Charles openly challenges him, Varney questions his sanity, all the while maintaining perfect manners.[84] Charles must admit defeat, as if giving up 'some contest in which he had been engaged'.[85]

Without realizing it, the other characters are drawn into a game of power at which the vampire is the most skilled player, disarming them with his civility and questioning their sanity when they come dangerously close to speaking the truth. The emphasis placed on the vampire's voice identifies him as the one who takes for himself the right to speak, the one whose words establish what is true, relegating all the other characters to the subordinate position of listeners. This dynamic conforms to other literary representations of vampires. Victorian fiction consistently constructed the vampire as a metaphor of power. Carol A. Senf notes that the Brontë sisters created their metaphorical vampires in the figures of

Catherine and Heathcliff in *Wuthering Heights*[86] and Bertha Rochester in *Jane Eyre*.[87] In Dickens's *Bleak House*, both Vhole and Krook are life-suckers, as the first feeds on Richard Carstone and the second on other people's secrets.[88] All these characters act as emotional vampires exercising psychological and social power over other characters.[89] The use of the vampire as a metaphor of social power, Senf maintains, comes from the very first representations of vampires in literature: Polidori's Lord Ruthven, LeFanu's Countess Carmilla Karnstein, Stoker's Count Dracula and, of course, Rymer's Sir Francis Varney.[90] Varney is an aristocrat, and therefore, Senf argues, a metaphor of social (tyrannical) power[91]; also, being 'a creature from the past who enters the present to influence it', he doubles as the power of the past over the present.[92]

Therefore, Varney's own nature as a vampire symbolizes exploitative power. His aristocratic role determines his social power, and his language shows the first signs of the psychological power exercised by vampiric characters in nineteenth-century fiction. There are grounds to suppose that, had the series not run on for so long, Varney would have turned out as cunning and malicious a villain as Sweeney Todd, who shows similar speech patterns and the same tendency to assert his power over other characters by distorting the truth while sadistically enjoying the process. However, while Todd is consistently evil, Varney's maliciousness tends to change in accord with his transformation from villain into rogue with a code of honour. I would argue, though, that the combination of his embodying a metaphor of power with his nature as a revenant offers another angle for interpreting his mastery of word manipulation and truth distortion. He, the displaced body par excellence, is reticent about his own identity, is the subject of medical experiments, and displays a general vengeful attitude towards humanity. He thus embodied the vengeful revenant that the characters of Mary Sinclair and Long Hannibal in *Manuscripts* expressed only in potential and served as an uncanny reminder of the countless bodies that were denied a peaceful rest for the sake of medical studies, still at the time in which *Varney* was being issued.

Opaqueness and secrecy were typical traits of the nineteenth-century medical world and extended to its language. Foucault notes that the development of the clinical gaze as the key to a privileged access to truth evolved into the idea of a 'speaking eye' that would be 'pure Language' and 'master of truth'.[93] This concept, he argues, favoured the revival of a certain 'medical esotericism': the language of the eye would be only understandable to the 'initiated', enacting a dynamic in which seeing the truth and knowing the language were mutual.[94] A consequence was that the medical community spoke an obscure language that concealed knowledge. Its utmost expression was the Anatomy Act itself. The Act displays several features that resemble the way in which Varney used language to establish relationships of power with the other characters. As only the vampire knew the actual meaning of his words, only the government, the medical fraternity, and workhouse masters knew the actual meaning of the Anatomy Act. In this way, they maintained a position of power over their counterparts, the other characters on the page and the poor in reality, who were baffled by obscure language and twisted sentence structure and found themselves involved in a challenge the rules of which escaped their comprehension. Exclusion from full understanding of the information and the code through which it was passed disempowered both the characters in the series and the readers of the Act. They were unable to control their own destiny, which, in the case of the pauper, included the disposal of their own remains.

Reticence and tendency to omit vital information characterizes the speech of Dr. Chillingworth. The narrative of his sensational revivification of Varney's remains enabled the author to indulge in grotesque and dramatic descriptions to his audience's satisfaction, but it also gave a glimpse of the doctor's mind and of how he shared or withheld information in a life and death situation. Ultimately, this reveals Chillingworth's views regarding anatomy and his attitude towards the dead body. From his story, it emerges that he considers obtaining corpses to dissect a 'desirable' and 'necessary' purpose.[95] Besides resembling Sir Astley Cooper's statements in front of the Select Committee, the concepts of 'desirability' and 'necessity' Chillingworth expresses were typical of such forms of pro-anatomy discourse as were being made available to the public, such as T.S. Smith's 'The Use of the Dead to the Living'. In the pamphlet, Smith defined the creation of legislation that would grant medical schools a steady supply of bodies for dissection a 'desired object'.[96] Similarly, he stated that the faculty of medicine and the Hôpital de la Pitié in Paris provided the pupils with 'all the facilities for dissection that can be desired'.[97] Alongside the idea of desirability, Smith endeavoured to stress the concept of 'necessity', particularly when referring to corpses. Parisian legislation, he wrote, provided the students with 'the bodies which are necessary for teaching anatomy'; a few lines below, he explained that 'the subjects that are necessary for teaching anatomy' were taken from the hospitals, and that practices and regulations of the city secured 'the necessary number of bodies'.[98] To support his argument, he tapped into the concerns regarding experimentation, claiming that, if the medical students were unable to access 'the necessary information' by dissecting the dead, they would have to acquire it by experimenting on the living poor.[99] He maintained that it was 'absolutely necessary' for surgeons to be acquainted with the internal organs of the human body,[100] and that a successful operation depended on the 'necessary' coexistence of proficiency in anatomical knowledge and command of the surgical tools.[101] Resurrectionists were included in this set of necessities, because they provided the 'subjects necessary for dissection'.[102]

Smith's emphasis on the necessity of providing more corpses for the study of morbid anatomy possibly originated in the concept of 'necessary inhumanity' formulated by Hunter, in which the knowledge of anatomy, and its subsequent dehumanization of the medical student, were essential to the instruction of a proficient surgeon. Dr. Chillingworth reiterates the image of the surgeon supporter of the necessity of dissection, while at the same time embodying the threat implicit in this ideology in the eyes of the wider public. The inclusion of pro-anatomy discourse in his speech would be particularly relevant to *Varney*'s working-class readership, who were powerless before the Act and perceived the medical fraternity as hostile, violent, and dangerous. These traits would necessarily be transmitted to the fictional doctor in the moment he proclaimed himself a supporter of dissection, which makes it unsurprising that his character is necrophiliac. His emphasis on the desirability and necessity of anatomy concretizes the wider public's anxieties about the idea of dissection—and indeed inhumanity—as necessary, and the conviction that this concept bespoke a dangerous lack of ethics. The new esotericism of the language of medicine, as Foucault defines it, provoked trust issues.

As mentioned in the Introduction chapter, Hurren notes that the Anatomy Act conveyed to the reader the impression that 'something was being withheld' regarding its true implications,[103] and Chillingworth's account of his resurrecting Varney conveys the same impression. We can consider the account from both the perspective of Chillingworth's reticence to sharing vital knowledge and what implicit information he does provide. The 'confession' the doctor makes to his friends has 'several times [...] trembled on [his] lips',[104] but he did not make it until events compelled him. His emphasis on his possessing information about 'him who has caused you so serious an amount of disturbance',[105] meaning

Varney, deflects attention from his own responsibility: his secrecy about his disturbing past practices jeopardized his neighbours. Obviously, he prioritized withdrawing information that might lead to questions on his professional ethic over the safety of his friends (and patients), which characterizes him as ambiguous and untrustworthy. He hid his knowledge of Varney's true nature, his acquaintance with the hangman, and most of all his role as the mad scientist whose hubris enabled the monster to hunt the family. He also hid his past neurosis so that his current authority could not be challenged. However, the information he provides in this regard allows the other characters, and the readers, to infer something about him as a practitioner.

Chillingworth describes the ideas of his younger self as 'mad-headed' and 'wild',[106] and his state as one of 'monomania'.[107] He claims that it was a direct result of his zealous pursuit of medical studies, which apparently conflated anatomy and dissection with lack of ethics. However, his narrative reveals that the real threat was not the study of medicine per se, but rather the unstable mind of the medical student Chillingworth, who approached a potentially destabilizing discipline with the wrong attitude from the beginning. Chillingworth's nervousness prevented him from benefiting from his medical training, from forming the rational, detached frame of mind typical of the proficient surgeon. The detachment he developed was imperfect, as it allowed him to experiment over human remains without qualms but not to see the truth, that is, the wider implications of his experiment and its potential risks. He can see no further than his own glorious dreams, which obfuscate and over-excite his mind. Therefore, his medical gaze is myopic, forever impaired by his delusions.

By contrast, Dr. North's clear-sightedness reflects on the power of his words. Whereas he is not entirely successful in fighting Varney—after all, a vampire is a figure associated with lore and superstition, while North is a man of science—he is the first to suggest that the family might be under the attack of a vampire. Notwithstanding his own scepticism, unlike Chillingworth, he immediately discusses his suspicions with the Croftons, preventing Varney from killing again. He can see truth more clearly than the other characters, and he acts conforming to what he sees. When he finds the marks of Varney's teeth on Clara's throat, he alerts the family; when Clara dies, he deliberately points to her corpse and tells Sir Crofton that 'that is not [his] child'. Not only does this statement invite the father to come to terms with the death of his daughter, but also contains a greater truth that the doctor, although he is not positive about it, may at least infer: Clara is now a vampire.

His privileged access to truth allows him to control information and endows him with the authority to lead the family through Clara's death. Dr. North chooses his interlocutors carefully. He manages to soothe Clara during the brief recovery preceding her death by drawing her into the secure space of the patient–doctor relationship. When Clara wakes up in a delirious state exclaiming 'I – am mad!', he invites her to 'tell [him] freely, as [her] medical man, what has happened', assuring her that 'if any human means can aid [her], [she] shall be aided'.[108] His words immediately soothe Clara; dialoguing with Dr. North, she becomes the patient, and takes the passive role of object to be gazed upon. It is up to 'her medical man' to use his superior knowledge to help her. After listening to her story, North deliberately proclaims her and her sister Emma, who heard her screams and was first on the scene, to be under the influence of nervousness and suggest they both take a rest together in the same room. However, when he is left alone with the other men, he asserts he does not believe the two girls to be delusional and formulates his theory about the vampire attack. Hence, while apparently acting in compliance with the Victorian habit, noted by Wood, of giving ideological biases precedence over the diagnosis, he regulates the distribution of knowledge according to what he believes appropriate for the benefit of the small community he is aiding. Granted, his behaviour is biased, as he supposes that Emma and Clara could not bear with hearing his theory on vampire attack; however, his biases only influence his policy of knowledge distribution, not his diagnosis. The peace of mind of his unnerved patients is paramount, as was the tranquillity of Long Hannibal's widow to the mature physician in *Manuscripts*. His withholding information, therefore, is the rational choice of a medical man acting in his patients' best interest.

Likewise, he deliberately excludes from knowledge all characters that he judges to be too weak to receive it. As the Crofton men develop the nervousness that had so far characterized the women of the family, the doctor progressively curtails communications with them. After he has taken care of lifting Clara's corpse from the floor and putting it back on the bed, Sir George asks him to explain to him the meaning of what is happening. Dr. North answers: 'Do not ask me [...] I cannot tell you; I confess I do not know what advice to give you, or indeed what to say to you'.[109] While this answer may simply be an admission of ignorance on the part of the doctor, and consequently a refusal to pronounce himself on what he does not know, the wording suggests something more. He 'cannot tell' what he thinks to Sir George, he does not know how to give him advice or how to talk to him, because Sir George does not represent a credible interlocutor for him. Not only does he not know the code through which North communicates—what Foucaultdefines the pure 'Language' of the medical gaze—but he also proved himself irrational and subject to his own passions. Dr. North's refusal to speak to Sir George may be actually a refusal to give him knowledge that North believes him unable to bear. Though unfair, it is the doctor's right to do so. Throughout the episodes that describe the attack upon Clara and her death, Dr. North's word is final because of his greater ability to 'see' the truth, which explains why his philosophic speech on death assuages the Crofton men. He is the authoritative repository of knowledge; hence his words are endowed with the status of truth and the other characters act in compliance with them.

Ironically, before Clara is attacked, the interlocutor Dr. North chooses is Varney. The narrative specifies that the vampire shows mastery of several scientific subjects, as well as 'no small amount of skill and theoretical information upon medical matters', to the extent that North allows him in the chamber of Clara's fiancé (who, as mentioned above, is recovering from a fall from his horse), while 'perhaps he would have objected to anyone else'.[110] The narrator comments on this decision declaring: 'Varney the vampyre could fascinate when he liked'.[111] Varney's charm mainly consists in using the correct code to communicate satisfactorily with the other characters, and he can speak the language of medicine. After all, he is a resurrected cadaver, the object that perhaps more than any other engaged in communication with the medical gaze, its key to truth. Varney is aware of Dr. North's power to penetrate truth. Perhaps, although unable to resist the temptation to engage in one of his covert battles for superiority with the most dangerous character in the family circle, he realizes that North represents a threat for him and makes sure to lock him into his room before attacking Clara. This shows he knows that the doctor, although he enjoys his conversation, cannot be fooled. Events will prove him right.

To conclude my discussion of truth and knowledge in *Varney*, it is necessary to turn, though briefly, on a third element, the mob, which the narrative examines in terms of its understanding—or, rather, lack thereof—of truth as related to the dead. The crowd as a disruptive element that provokes chaos is present in both the Bannerworth and the Crofton saga. In one of the last episodes of the series, they destroy Clara in her vampire form. However, it is in the Bannerworth saga that the mob is explored at length as an element ultimately incapable of 'seeing', and therefore accessing, the truth, an inability which is ascribed chiefly to the ignorance of the individuals in the mob and its collective expression.

In Chapters 44 and 45, Sir Francis Varney' presence drives the neighbourhood into a state of vampire hysteria.[112] Townspeople suspect that the spawns of the vampire will soon attack them. 'There was Miles, the butcher', they say, 'you know how fat he was – and then how fat he wasn't'.[113] They unanimously decide to storm the cemetery and disinter Miles the butcher. The mob in *Varney* is generally a brutal, blind, and dumb entity, subject to sudden and terrible bursts of rage. This mob, however, is particularly stolid and gullible, as they fall for the words of a fanatic preacher, Mr. Fletcher; as we have seen in *Manuscripts*, in Rymer's fiction this is an unpardonable sin, as it was in the fiction of many Victorian writers, most notably Dickens. During the scene, Mr. Fletcher works up the enraged populace with a speech in which he compares them to 'vampyres […] men who walk in the darkness when the sunlight invites [them]', who 'listen […] to the words of humanity when those of a diviner origin are offered to [them]'.[114] The agenda behind his solemn speech becomes apparent when he states that, whatever discovery they make by opening the coffin, the result shall be the same: if they find the butcher 'decaying … [they] shall gather from that a great omen, and a sign that if [they] follow [him] [they] seek the Lord'; should they find him 'looking fresh and healthy' it will be likewise a sign 'that what [he] say[s] to [them] shall be the Gospel, and that by coming to the chapel of the Little Boozlehum, [they] shall achieve great salvation'.[115] In brief, Mr. Fletcher presents himself as the depository and dispenser of the words 'of a diviner origin' that should enlighten them. Moreover, the self-interested preacher sanctions the desecration of a tomb with his speech. He provokes a disruption of the cycle of life and death for his own personal gain,[116] manipulating the mob, exploiting their superstitious fears. His speech is based on obviously shaky premises, but it nonetheless incites the mob to violate the religious values of the community. It is unsurprising, therefore, that the much advertised 'sunlight' should fail to manifest itself to the crowd, as the opening of the tomb of Miles the butcher brings them no closer to the truth.

Upon opening the coffin, they find its only occupant is, quite inexplicably, a brick. The narrative does not elaborate on the origins of such interesting phenomenon, whereas it focuses on the crowd's behaviour before and after the discovery, arguing that the mob lacks the intellectual means to achieve any useful knowledge from the observation of death. While the narrator acknowledges that human nature shows 'a natural craving curiosity' towards the dead, it holds that only a certain type of individual is able to benefit from it, namely those men 'of education and endowment' who would undertake long journeys to attend the exhumation 'of someone famous in his time', in which case curiosity takes a 'refined and sublime shape'.[117] To this figure, the narrator juxtaposes 'the vulgar and ignorant', who display the same sentiment in a 'grosser and meaner form', conforming 'with their habits and thoughts'.[118] The narrator may or may not have alluded specifically to doctors and their refined education when eulogizing the first set of men's attitude, while there is no mistaking his certainty of the mob's ignorance, their inability to use their own gaze to reach simple truths, and their spite, in fact, for whoever attempts to do so. The only person in the mob who tries to think logically is a boy, who is rebuked and receives several boxes in the ears for asking questions. The boy juxtaposes to Dick, the man the mob elects as its spokesperson, who is characterized by his repeated unsuccessful attempts at using scientific language. While the boy asks sensible, evidence-based questions (for instance, he asks how can the butcher possibly leave a coffin that lays so heavily under the mud, and receives a cuff for not minding his business), Dick accuses him in rapid sequence of ignorance and excessive curiosity. This fight between sight-based curiosity and blind ignorance reaches its apex when the boy asks doubtfully, looking at the brick: 'you don't mean to say that's the butcher, Dick?' Dick taps him on the head with the brick and exclaims 'There [...] that's what I calls occular [sic] demonstration. Do you believe it now, you blessed infidel?'.[119] In one move, Dick shows that he does not know what an ocular demonstration is, and confuses science with blind belief. He goes so far as to predict the death of the boy, stating that he is such an 'inquiring genius' that he will 'get [his] head in some hole one day, and not be able to get it out again and then [he, Dick] shall see [the boy] a kicking'.[120] This prophecy, which might be simply picturing a ridiculous scene, as well as alluding to the image of a body hanging from the executioner's noose, indicates Dick's spite for curiosity.

The cemetery scene in the Bannerworth saga, therefore, is a précis of how the narrative portrays the approach to knowledge. Characters such as Mr. Fletcher cannot pass any type of knowledge, because their words are false, informed as they are on the individual's personal agenda. It must be noted that Mr. Fletcher is duly rewarded for his pretence of truth: as the mob enthusiastically sets to disinter Miles, Mr. Fletcher's 'rather [...] capacious' mouth is filled with a 'descending lump of mould of clayey consistency', which almost causes the preacher to take out his own teeth in the attempt to remove it.[121] The mob and its single individuals are likewise excluded from the truth: their falling for the preacher's duplicitous words makes their blindness apparent. They do not know the language of science because they are unable to use the chief organ of scientific investigation, the gaze. While the character of the boy suggests that even an untrained gaze, if used properly, can lead to truth, the people in the mob cannot exercise their gaze, and hence their discernment. When their eyes show them an empty coffin and a brick, they accept Dick's makeshift explanation that a 'transmogrification' must have happened that made Miles 'consolidify' himself in a brick.[122] Notably, neither the characters nor the narrator considers the chance that it might have been the action of bodysnatchers; the core point is that the mob does not possess the intellectual means to access truth, as opposed to the men of education whose curiosity towards death is acceptable.

The mob episode synthetizes the way the narrative engages with processes of access to, and transmission of, knowledge, praising sight-centred curiosity and education, which are the only means to reach the truth, as it is extant in the character of Dr. North. In *Varney*, the gaze is the supreme channel of truth, in a paradigm that tallies with Foucault's theorization of the speaking eye as pure Language in nineteenth-century medical discourse.[123] When displaced bodies appear around medical men, the language of knowledge in *Varney* is the language of science, in which the trained eye, master of truth, creates the code, constructing the new medical esotericism Foucault describes. Chillingworth represents the distortion of its principle. His speaking eye is myopic, impaired as it is by his lack of rationality, and instead of policing the distribution of truth wisely, he jealously keeps vital truth s to himself for fear of antagonizing the other characters, ultimately jeopardizing them. North, by contrast, embodies the positive clinical gaze. While he achieves and maintains a position of power thanks to the privileged access to knowledge his trained gaze allows him, he is a reliable man who speaks the truth and controls the distribution of knowledge in everyone's best interest. Unlike his ancestor, he is a healer, and represents the new medical science, enlightened and enlightening. He speaks the language of science, which is the language of truth par excellence, and he has a positive influence on the other characters, particularly the male ones. The mob's ignorance, instead, prevents them from seeing the truth at any level: their inability to exercise their gaze shows that blindness, which stands for ignorance, is incompatible with the language of science.

3 Life–Death Inversion and the Gaze: Disruptive Effects of the Untrained Gaze on the Corpse

If the mob's blindness prevents them from accessing the truth, its impaired gaze nonetheless focuses—or attempts to focus—on the dead body. The narrative frames the medical man's gaze as the source of truth; it follows that the act of gazing over death plays a crucial role in the plot. Not only does the series discuss the appropriate application of the medical gaze, but it also explores the consequences of applying a defective gaze, that is, an untrained gaze, on death.

When the uncanny disruption of the life–death cycle occurs, the gaze of the untrained individual usually fixes over the displaced body. Naturally, this is more likely to happen in the spaces of the cemetery and the vault. As in *Manuscripts*, these are peculiarly restless spaces, in which tombs are discovered to be empty, corpses are found out of place, and where the supernatural presence of the vampire ascribes a literal meaning to the word 'resurrection'. Unlike in *Manuscripts*, by contrast, the uncanny inversion from death to life yields no positive outcome: it is always synonym of chaos and disruption, either in the form of a violation of the values underpinning the relationship between death and society, or in the form of the vampire's supernatural agency in the world of the living. The gaze of the untrained observer reflects the inversion's intrinsic uncanniness, as it is either inappropriate, or horrified. Voyeurism is a crucial component of the untrained, defective gaze on the corpse, underpinning the negative results it produces.

Two instances of this type of gaze are instrumental in illustrating this point: the gaze of the mob on Miles's coffin in the Bannerworth saga, and the gaze of Will Stephens, the sexton, on Clara's displaced body in the Crofton saga.

In the case of the mob in the cemetery, the narrative frames morbid curiosity as the propeller of the event, which it states is natural in the human being.[124] However, the narrator's tone remains undisguisedly derogatory as the mob digs up Miles's coffin, and indulges in the gory and repulsive aspects of their behaviour. As the spade touches the coffin's lid, the excitement of the crowd increases: the people near the brink of the grave '[crouch] down' to look at the coffin, ignoring 'the damp earth that [falls] upon them' and 'the frail brittle and humid remains of humanity that occasionally [roll] to their feet'; the scene, the narrator asserts, 'only want[s] a few prominent features in its foreground of a more intellectual and higher cast than composed by the mob, to make it fit for a painter of the highest talent'.[125]

Perhaps, the narrator is referring to such Hogarthian works as 'The Idle Apprentice Executed at Tyburn' from *Industry and Idleness* (1747), or 'Noon' from *The Four Times of the Day* (1738). The emphasis placed on the 'frail brittle of humanity' seems planned to raise disgust, and I would venture that the narrator's attempt at stirring the disgust of the reader before the voyeuristic excitement of the mob, combined with the allusion to a visual representation of the scene, suggests that the author might have been drawing inspiration from real events. Though it is impossible to make any conclusive assertions as to whether Rymer was ever a spectator of such a scene, desecration of burial grounds at the hands of unruly mobs was not unlikely in nineteenth-century London. As we have seen, the upturning of graves was common during cholera outbreaks, and the press covered such events. Moreover, in direct connection with the displacement of cadavers, mobs figured also in newspaper reports of resurrectionist apprehensions. An issue of the *Times* from 1832—soon after the passing of the Act—reports the arrest of a gang of resurrectionists in Hereford. As the prisoners were escorted to the Mayor's house for interrogation, 'a large mass of persons' gathered 'anxiously waiting' the verdict, 'but still more anxious to catch a glimpse of the resurrectionists', as though they believed that bodysnatchers 'must present a different appearance from their fellow-men'.[126] When the prisoners were moved to the county gaol, 'an immense crowd' followed them and 'assailed with the most vociferous expressions of disgust and contempt'.[127] Four months later, a similar scene happened in Kent, where three men were found with two bodies packed on a cart. The mob in this case was even more violent, because of the suspicion that the three men might be burkers. The crowd that quickly assembled around the station house was allegedly composed 'of several thousand persons' and called for the police to surrender the 'Burkites' 'in the most menacing and outrageous manner'.[128] Such was the intensity of the agitation of the mob, the reporter claimed, that for a moment 'it was thought that the station-house would be completely pulled down' and as the men were escorted outside by the police, the crowd attacked them 'with stones, bricks, and missiles of every description'; the police apparently had 'the utmost difficulty to prevent their prisoners being sacrificed by the indignant multitude'.[129]

Both articles strive to frame the mob as dangerous. The journalists describe the crowd as uncannily vast—'several thousands', 'large mass', 'immense crowd'—and in a state verging on hysteria. They also emphasize its disregard for authority, as they follow and/or attack the police. The officers appear helpless before the enraged mob, who in turn is (allegedly) capable of almost destroying the station house. The term 'sacrifice' applied to the prisoners, which is suggestive of tribal behaviour, implies a regression of the mob to a savage state.[130] This portrait is unsurprising, as radicalism, and the political turmoil this provoked, made the Victorians profoundly distrustful of large crowds, a distrust that also emerges, for instance, in Dickens's writing. In *Oliver Twist*, as Oliver runs for his life from the frenzied crowd that accuses him of pick-pocketing, Dickens comments on the 'passion for *hunting something*' that characterizes the human being.[131] The comment suggests that the individual regresses to a primeval, savage state, as much as the cemetery scene in *Varney* expresses the mob's transgression of the cultural norms that regulated the relationship between the community and its dead. The threat posed by the disruption of boundaries between the surface and the underground, with the subsequent invasion of the world of the living by dead matter, is identified with a regression from civilization to savagery.

Inverting the regression is difficult. As the superstitious fear of the dead starts worming its way into the collective mind of the mob, their determination falters. They try to reassure themselves that they are not, in fact, committing a sacrilege, but exercising their right to access the truth: 'if he's a vampyre, we ought to know it; if he ain't we can't do any hurt to a dead man'.[132] They even wonder whether they should read the service for the dead, at which point the boy, the voice of reason, mocks them: 'Yes [...] I think we ought to have that read, back-wards'.[133] This blasphemous suggestion exposes the pretence of religious feeling the mob is endeavouring to sustain, triggering a rush of hypocritical indignation that deflects attention from the actual desecration that is being committed and from the voyeuristic curiosity that propelled it.

The disruption of the rest of the dead in the space of the cemetery in *Varney*, therefore, could not possibly yield positive results, because it is founded on superstition and on the voyeuristic exercise of an untrained gaze. Unlike in *Manuscripts*, where the clinician's gaze transforms the space of death into the space of life, the collective defective gaze of the unruly mob in *Varney* transforms the space of the cemetery in the ultimate space of chaos. Drawing from contemporary representations of the disruption frenzied crowds provoked in Britain's geospace, which tap into deeper anxieties about popular commotion, the narrative presents the results of the uncontrolled application of the untrained gaze into the space of death as a mark of savagery. It is not surprising, therefore, that the mob is denied the sight they so craved, and that all they can see in the violated tomb is a brick.

The episode concerning Clara Crofton's resurrection as a vampire merges these points with a supernatural component which suggests that not only does the untrained gaze yield disruptive results, but the contemplated object itself may impact negatively on the untrained eye. The inverted movement is represented through the dead literally ascending from the space of the tomb, curtailing the distance between death and the living in a way that overwhelms the beholder. Proximity to death is not for the layperson's gaze, as suggests the character of Will Stephens, the sexton of the village in which the Croftons live.

The sexton, is, unsurprisingly, a greedy, untrustworthy character. As mentioned in the previous chapter, sextons were often associated with bodysnatchers and body displacement. The anonymous bodysnatcher A.B. was very clear on this point with the Select Committee for Anatomy, declaring that if a bodysnatcher meant to 'get subjects for any constancy out of any burying-ground', the only way to do it was by bribing the staff.[134] The one thing that could dissuade a sexton or a gravedigger from consorting with resurrectionists, A.B. added, was the danger of losing their job. If the current gravedigger, sexton, or guardian was replacing a man who had been fired after his partnership with resurrectionists was discovered, he would categorically refuse bribes.[135] Although the narrative does not clarify whether Will Stephens is in league with resurrectionists, he fits the dodgy sexton stereotype. Sir George Crofton paid him to put sawdust in Clara's vault, but Will pockets most of the money and purchases a small amount of sawdust with what is left. Moreover, he stops for a few drinks before attending to the task, which brings him to the church at the appropriately solemn time of midnight, on an appropriately dismal night, conforming to the Gothic tone of the series and to the melodramatic style of the penny bloods. The first instalment relating Will's adventure in the tomb concludes abruptly on Will nursing his beer and a strange dark figure, apparently endowed with superhuman strength, forcing its way into the church. The narrator prepares the reader for what is to come by announcing imminent frightful events 'that require the closest attention' and will happen 'in the vaults'.[136] He specifies that they are guaranteed 'to fill the reflective mind with the most painful images, and awake the sensations of horror at the idea that such things [...] are permitted tacitly by Heaven to take place on the beautiful earth destined for the dwelling place of a man'.[137]

Such a melodramatic hint, albeit not leaving much to the readers' imagination, was guaranteed to secure their attention and, naturally, their purchase of the next instalment. A detail, however, that is worthy of closer attention is the emphasis placed on the spatial frame of the event, the vault, and the statement that what will happen there will affect the 'dwelling place of the man', that is, the world of the living. The event is framed as something opposed to the idea of 'Heaven', hence of holiness, of sacred; it will be, in brief, a distortion. The narrator, of course, is alluding to Clara's resurrection as a vampire. As the emphasis on the space of the vault hints, the ensuing episode focalizes on the absence of the corpse from the coffin, and indeed from the vault altogether. After a few humorous accidents, including fancying himself murdered when he trips and falls on the bag of sawdust, Will finally descends into the Crofton family's vault. Here, he spots a coffin lid at his feet, which he recognizes as that of Clara's coffin. Will immediately starts 'trembling and turning over in his mind all the most frightful explanations of [sic] what he saw [...] 'Has she been buried alive? Have the body snatchers been after her? [...]''.[138] It is curious that resurrectionists should appear in *Varney* only at this point, while they would have explained very well the disappearance of Miles the butcher.[139] Possibly, the presence of a sexton, a figure commonly associated with the resurrectionists, favoured the introduction of the subject. Gazing over the misplaced coffin lid, Will contemplates 'all' the possible frightful explanations for the situation, which, in the end, are two: either Clara has been buried alive, or she has been stolen by bodysnatchers.

It is hard to imagine that someone, even a stronger person than the delicate Clara Crofton, could awaken from a coma and, in the agony of suffocation, force open the nailed lid of a coffin; yet, the idea of premature burial, as we have seen in the previous chapter, alarmed the Victorian public and fascinated authors.[140] Will resolves to ascertain whether the body is still there, which would mean that 'she had been buried alive, and had just strength enough to force open the coffin [...] and then to die in that horrible place'.[141] However, when he looks into the coffin, it is empty; the sexton stares at the empty casket 'as if there was something peculiarly fascinating in it, and most attractive, and yet, nothing was in it, no vestige even of the vestments of the dead'.[142] Besides focusing on yet another character fascinated by the contents—or lack thereof—of a coffin, the emphasis on the casket's emptiness indicates that something is amiss. The absence of burial clothing excludes immediately the possibility of resurrectionists, as they only took the naked body and left the shroud in the coffin, to avoid being arrested for theft. The idea of the supernatural starts forming in Will's mind and he suddenly becomes aware that he is inside a tomb with an open, empty coffin. The living sexton fears the ghostly presence of the dead girl; the empty coffin suggests that he may be doomed to replace the dead in the vault. He clumsily dashes for the stairs and manages to make it to the floor above, where he finds, to his great relief, 'that the night has turned out so fair and beautiful'.[143] Ascending from under the earth, the space of the dead, to the light above the ground is of immediate relief to the living man. Yet, Will soon discovers that he is not the only one who left the space of the tomb, and that the dead are walking on the space above. Perceiving someone else's presence in the church, he climbs to the gallery, to be at once in a safer position and in a better spot to survey the church. From there, he hears Varney's voice beckoning Clara, exhorting her to awaken, and automatically directs his gaze towards the sound, that is, downwards. In the space below, 'in a pew just beneath him', he spots a human form 'lying in a strange huddled up position'; the moonbeams fall upon it, and to the 'experienced' eyes of the sexton, it appears clear that 'it [is] arrayed in the vestments of the dead'.[144] Will Stephens's eyes are used to the sight of the dead; they are by no means prepared, however, to cope with the displacement of a body and the subsequent disruption of the safe distance between him, a living person, and dead matter. There are two points to make in this regard, and they both pivot on the fact that Will's gaze upon Clara's corpse is out of place in the church. Firstly, Will's point of view, looking down from a higher position on a pew with a corpse stretched over it, resembles the gaze of the medical student in the operating theatre: the gaze of the observer is directed downwards, from an elevated position, onto the operating table where the subject lays. The sexton, however, is no clinician, and the position he is maintaining is not appropriate for his gaze. Secondly, the Victorians had their methods to satisfy their voyeuristic curiosity towards death while at the same time maintaining a safe distance from it, and they all involved the staging of death. Staging granted safety boundaries, which Will's position breaks.

The staging of death pervaded nineteenth-century popular culture. Besides the gruesome spectacle of hangings, and, earlier in the century, public dissections, staging death also related to what Hurren terms 'the dissection room drama'.[145] This expression signifies the sensational, dramatized representations of dissection rooms in the press, which the medical community used to advertise medical studies among the middle class.[146] The public dimension of the staged display of the dead, instead, took the form of the anatomy museums, which allowed the layperson to behold, that is, to observe, a sanitized version of the results of dissection.[147] The anatomical Venus, examined in the previous chapter, was part of this staging, which, Hurren argues, allowed the viewer to engage with the actual dismemberment of the body from a safe distance.[148] Interest towards the staged display of the dead finds confirmation also in the British accounts of the Parisian morgue, where the eager public—including the British tourist—could see the bodies of unclaimed suicides or homicide victims.[149] While criticizing the barbaric foreign custom of showing the dead, the British onlooker could indulge their own voyeurism, which emerged in the minutely detailed, lurid descriptions of the travellers.[150] On both sides of the Channel, death and the dead were, more or less overtly, a show to look upon; the staging removed the onlookers to the position of spectators, not directly involved in the drama. Paul Vita considers the traveller's report as the ultimate removal operation, which turns the show, and therefore death, into 'a matter of vicarious readings, not personal experience'.[151]

The reader's vicarious experience of Clara Crofton's resurrection differs from the account of morgue experiences: through Will's gaze on Clara's dead body stretched on the pew, the reader experiences not the satisfied gaze of the voyeuristic morgue visitor, but the dread of the sexton beholding the displaced corpse. Clara's body lacks the stillness characterizing corpses exposed in the morgue, the museum specimens, and the recumbent, sensual anatomical Venus. Stillness and the subsequent certainty that the spatial distance between the observer and the dead body would be preserved were crucial to guarantee that the experience would remain, to use Vita's words, 'essentially visual'.[152] Clara's corpse, instead, moves from its place in the tomb towards the sexton, curtailing the distance between the living and the dead. The uncanniness of this proximity provokes in the sexton a terror the readers vicariously experience, finding themselves alarmingly close to the corpse. Will repairs to an elevated position to regain his role as a distant observer when he feels uncomfortably close to the dead; Clara's resurrection as a vampire again annihilates the distance. The movement turns the sexton—and, by extension, the reader—from a removed spectator into an involved, but powerless, observer. Exercising the untrained gaze over the displaced corpse, therefore, only provokes terror. The untrained eye perceives the proximity of death as a threat. The space of the vault in *Varney* and the disruptive movement of its contents expose the voyeuristic lay-eye of the reader to the sight of the displaced corpse, provoking the same terror raised in the sexton.

The supernatural component the vampire introduces disrupts the plots' rationality and transforms the space of the cemetery and the vault in spaces of un-rest, chaos, and anxiety, as cemeteries indeed were in the nineteenth century. The restlessness characterizing the geospace of the cemetery surfaces in the disruptive inversion from death to life that characterizes the cemetery/vault narrative space, which in turn highlights the relationship between the untrained gaze and the dead body. The relationship is reciprocally disruptive: on the one hand, the voyeuristic gaze of the mob dehumanizes the observer while desecrating the observed object; on the other hand, the sudden proximity of death horrifies the untrained gaze of the sexton, who is unable to regain his position of distant observer and is overwhelmed by the experience. This representation, I argue, admonishes the voyeuristic gaze of the readers themselves: as the average purveyor of penny blood series would be driven by a high degree of voyeurism, the narrative suggests that the pursuit of the 'show', the staged spectacle of death, may lead the untrained eye to survey sights with which it is not prepared to bear. The juxtaposition of the mob's voyeurism to the educated man's aesthetic curiosity and more practically, the staging of Will Stephen's gaze over Clara's body as a 'dissection room drama' imply that the gaze over the dead body is appropriate only in specific spaces, namely the spaces of medicine. When removed to different spaces, it is not guaranteed that the superstition and ignorance that characterize the untrained gaze will not yield unexpected or unpleasant results.

Varney received damning criticism from scholars in its afterlife as a remnant of the penny blood era. In 1998, Richard Davenport-Hines labelled it 'unreadable'.[153] Yet, as I have shown, this long and convoluted narrative engaged, as regards medical history, with very modern concepts that characterized the evolution of the discipline and how it was perceived by the public. Granted, the style was not overly refined, and this typically melodramatic narrative did exploit its audience's voyeuristic taste for gore and thrill using such Gothic features as the Frankenstein story and the element of the supernatural. However, in *Varney*, these components build a rich discourse that includes several important aspects that Foucault and medical historians later individuated as pivotal to nineteenth-century medical discourse.

The exploration of the medical gaze and its language through the figures of the two doctors relates to the trust issues and power struggles that characterized the relationship between the medical fraternity and the public, more specifically the working-class. The series also portrays the pervasiveness of the medical discourse in other discourses, such as masculinity. This shows awareness, on the part of the narrative, not only of the existence of such pervasiveness, but also of its role in the unfolding of power dynamics within society. Finally, in a historical moment in which the appropriateness of the medical gaze on the dead body was being discussed and evaluated, *Varney* questions the appropriateness of the untrained gaze over the same object, interrogating its motives. To sum up, *Varney the Vampyre* contributed to circulating the medical discourse(s) of its age, particularly the debate around the ethics of medical education, among a category of people who were excluded from the conversation on medical education while being, to a meaningful extent, its object.

Issues of inclusion and exclusion from certain discourses, as well as the question of the access to truth and circulation of knowledge, are explored in relation to the dismemberment and disposal of remains in one of the most famous narratives belonging to the penny blood genre. In Sweeney Todd's tonsorial parlour, with its ghastly mechanical chair and its secret passage to the basement of Lovett's pie shop, control over the spoken word is crucial to the thriving of the business.

Notes

1. With this term, literary critics label sequences of episodes in *Varney* that revolve around a specific set of characters, such as the Bannerworths or the Croftons.
2. Marmaduke Bannerworth's actual identity is unclear. At first, he is the man in the portrait, who was identified as Runnagate Bannerworth, a debauched ancestor; then, he becomes Henry and Flora's father. In James Malcolm Rymer, *Varney the Vampyre; or: The Feast of Blood*, ed. Curtis Herr, 2008th ed. (Crestline, CA: Zittaw Press, n.d.), 67, Curtis Herr ascribes this and similar oddities to the fact that, soon after he started writing *Varney*, Rymer began working on nine more series.
3. James Malcolm Rymer and Thomas Peckett Prest, *Varney the Vampyre; or: The Feast of Blood*, ed. E.F. Bleiler (New York: Dover Publications, 1973), 1:330.
4. Rymer, *Varney the Vampyre; or: The Feast of Blood*, 310.
5. Rymer and Peckett Prest, *Varney the Vampyre; or: The Feast of Blood*, 1:285.
6. 'monster, n., adv., and adj.' *OED Online*. Oxford University Press, March 2016. Web. 30 March 2016, def. 5.
7. Ibid., 1:301. My emphasis.
8. Chris Baldick, *In Frankenstein's Shadow—Myth Monstrosity, and Nineteenth-Century Writing* (Oxford: Clarendon Press, 1987), 141–42.
9. Ibid., 142.
10. Ibid.
11. Rymer, *Varney the Vampyre; or: The Feast of Blood*, 309.
12. Rymer and Peckett Prest, *Varney the Vampyre; or: The Feast of Blood*, 1:81.
13. Ibid., 1:331.
14. Ibid., 2:844.
15. Ibid., 1:326.
16. Ibid., 1:327.
17. Ibid.
18. Elizabeth T. Hurren, *Dying for Victorian Medicine—English Anatomy and Its Trade in the Dead Poor, c. 1834–1929* (London: Palgrave Macmillan, 2012), 82–84.
19. L.J. Grainer, 'A Medical Student in Search of a Supper,' *The Penny Satirist*, no. 164 (1840): 3–4. Quoted in Hurren, *Dying for Victorian Medicine*, 83.
20. Ibid., 84.
21. Ibid., 83.
22. Rymer and Peckett Prest, *Varney the Vampyre; or: The Feast of Blood*, 1:327.

23. Ibid., 1:328.
24. Ibid.
25. Pamela K. Gilbert, 'Sensation Fiction and the Medical Context,' in *The Cambridge Companion to Sensation Fiction*, ed. Andrew Mangham (Cambridge: Cambridge University Press, 2013), 189.
26. See Tabitha Sparks, *The Doctor in the Victorian Novel* (Farnham: Ashgate, 2009), 12–13.
27. Rymer and Peckett Prest, *Varney the Vampyre; or: The Feast of Blood*, 1:327–28.
28. 'Report from the Select Committee for Anatomy' (London, 1828), 18. Like Chillingworth and Frankenstein, Sir Astley also created his personal Monster, that is, Ben Crouch, the leader of the 'regular' gang of bodysnatchers in early nineteenth-century London. In *Life of Sir Astley Cooper*, Crouch is described as a sort of hulking underworld dandy, a pock-marked prize-fighter. Seldom drunk, although 'most abusive and domineering' when so, he made sure to be the only sober member of the gang when it was time to give each man his share of the payment, so that he could cheat the others out of their money. Bransby B. Cooper, *The Life of Sir Astley Cooper, Bart, Vol. I* (London: John W. Parker, West Strand, 1843), 1:413. In Ruth Richardson, *Death, Dissection and the Destitute* (Chicago and London: The University of Chicago Press, 1987), 71, Richardson states that Crouch and Cooper 'cordially hated' each other. She suggests that Sir Astley was thinking of Crouch when he stated in front of the Committee that resurrection men were the 'lowest dregs of degradation' ('Report from the Select Committee for Anatomy,' 18), and argues that Cooper's testimony 'aimed, in the long term, at denying Crouch and his ilk their livelihood'. Yet, she asserts, if truly Cooper introduced Crouch to the profession, then he was partially responsible for Crouch's corruption. Richardson, *Death, Dissection and the Destitute*, 71.
29. Rymer and Peckett Prest, *Varney the Vampyre; or: The Feast of Blood*, 1:328.
30. Ibid.
31. Rymer, *Varney the Vampyre; or: The Feast of Blood*, 310.
32. Rymer and Peckett Prest, *Varney the Vampyre; or: The Feast of Blood*, 1:328–29.
33. Ibid., 1:330.
34. Ibid.
35. Ibid.
36. Ibid.
37. Emma Liggins, 'The Medical Gaze and the Female Corpse: Looking at Bodies in Mary Shelley's Frankenstein,' *Studies in the Novel* 32, no. 2 (2000): 133.

38. Rymer and Peckett Prest, *Varney the Vampyre; or: The Feast of Blood*, 1:330.
39. Richardson, *Death, Dissection and the Destitute*, 43–44.
40. Ibid.
41. Ibid., 44.
42. Ibid. For more information on the poor and hospitals, see also ibid., 42–50.
43. Rymer, *Manuscripts from the Diary of a Physician, 2 Vols.*, 1:4.
44. Rymer and Peckett Prest, *Varney the Vampyre; or: The Feast of Blood*, 1:328.
45. Hurren, *Dying for Victorian Medicine*, 83.
46. Rymer and Peckett Prest, *Varney the Vampyre; or: The Feast of Blood*, 1:329.
47. Ibid.
48. Ibid., 1:330.
49. Gilbert, 'Sensation Fiction and the Medical Context,' 186.
50. Liggins, 'The Medical Gaze and the Female Corpse: Looking at Bodies in Mary Shelley's Frankenstein,' 132.
51. Ibid.
52. Ibid.
53. Rymer and Peckett Prest, *Varney the Vampyre; or: The Feast of Blood*, 2:797.
54. Ibid.
55. Ibid.
56. Ibid., 2:798.
57. Ibid.
58. Quoted in Hurren, *Dying for Victorian Medicine*, 205.
59. Quoted in ibid., 206.
60. Jane Wood, *Passion and Pathology in Victorian Fiction* (Oxford: Oxford University Press, 2001), 60.
61. Ibid., 64.
62. Ibid., 60.
63. Ibid., 64.
64. Rymer, *Varney the Vampyre; or: The Feast of Blood*, 728.
65. Rymer and Peckett Prest, *Varney the Vampyre; or: The Feast of Blood*, 2:798.
66. Ibid.
67. Ibid.
68. Ibid., 2:804.
69. Michel Foucault, *The Birth of the Clinic* (London: Routledge, 2003), 109.

70. Tabitha Sparks, 'Surgical Injury and Narrative Cure in Wilkie Collins's Poor Miss Finch and Heart and Science,' *Journal of Narrative Theory* 32, no. 1 (2002): 14.
71. Terrie M. Romano, 'Gentlemanly Versus Scientific Ideals: John Burdson Sanderson, Medical Education, and the Failure of the Oxford School of Physiology,' *Bulletin of the History of Medicine* 71, no. 2 (1997): 228–29.
72. See Sparks, 'Surgical Injury and Narrative Cure,' 14–22; Romano, 'Gentlemanly Versus Scientific Ideals,' 245–46.
73. Richardson, *Death, Dissection and the Destitute*, 44.
74. Hurren, *Dying for Victorian Medicine*, 82–83.
75. Wood, *Passion and Pathology in Victorian Fiction*, 64.
76. Romano, 'Gentlemanly Versus Scientific Ideals,' 227–29.
77. Liggins, 'The Medical Gaze and the Female Corpse,' 133; Sparks, 'Surgical Injury and Narrative Cure,' 2.
78. Rymer and Peckett Prest, *Varney the Vampyre; or: The Feast of Blood*, 1:61.
79. Ibid., 1:63.
80. See Senf, *The Vampire in 19th Century English Literature*, 34.
81. Rymer and Peckett Prest, *Varney the Vampyre; or: The Feast of Blood*, 1:76.
82. Ibid., 1:78.
83. Ibid., 1:79.
84. Ibid., 1:80.
85. Ibid., 1:79.
86. Senf, *The Vampire in 19th Century English Literature*, 75–93.
87. Ibid., 94–100.
88. Ibid., 104–20.
89. Senf notes that the possibility of Catherine, Heathcliff, and Bertha being actual vampires is never explicitly denied in the narratives.
90. Ibid., 43.
91. Ibid.
92. Ibid., 44.
93. Foucault, *The Birth of the Clinic*, 141.
94. Ibid.
95. Rymer and Peckett Prest, *Varney the Vampyre; or: The Feast of Blood*, 1:328.
96. Thomas Southwood Smith, *The Use of the Dead to the Living* (London: Baldwin and Cradock, 1828), 50.
97. Ibid., 51.
98. Ibid., 50.
99. Ibid., 53.

100. Ibid., 4.
101. Ibid., 18.
102. Ibid., 33.
103. Hurren, *Dying for Victorian Medicine*, 28.
104. Rymer and Peckett Prest, *Varney the Vampyre; or: The Feast of Blood*, 1:327.
105. Ibid.
106. Ibid., 1:330.
107. Ibid., 1:328.
108. Ibid., 2:795–96.
109. Ibid., 2:804.
110. Ibid., 2:792.
111. Ibid., 2:792.
112. Vampire hysteria was a phenomenon that sparked in the 1600s–1700s eastern and central Europe and consisted in the belief that actual vampires were spreading, killing, provoking epidemics, and spawning more vampires. On suspicion of vampire activity, crowds would storm cemeteries and dig up corpses to ritually kill the 'vampires'. See Senf, *The Vampire in 19th Century English Literature*, 20; Twitchell, *The Living Dead: A Study of the Vampire in Romantic Literature*, 18–19.
113. Rymer and Peckett Prest, *Varney the Vampyre; or: The Feast of Blood*, 1:202 During the vampire plague, several ordinary diseases such as porphyria, pernicious anaemia, tuberculosis, cholera, and even cancer were believed to be symptom of vampirism. The sentence implies that Miles the butcher had wasted away, which suggests that he might rather have died of tuberculosis or cancer, and also that Rymer was aware of vampire folklore. See Twitchell, *The Living Dead*, 19.
114. Rymer and Peckett Prest, *Varney the Vampyre; or: The Feast of Blood*, 1:205.
115. Ibid., 1:205–6.
116. Londoners were not unfamiliar with this type of dynamic. A famous case in which a man of the church sanctioned the disinterment of corpses to speculate over burial fees was that of Enon Chapel, covered by the press and brought to public attention by Dr. George Walker. The case will be examined in more detail in the next chapter.
117. Ibid., 1:203.
118. Ibid.
119. Ibid., 1:208.
120. Ibid., 1:210.
121. Ibid., 1:206.
122. Ibid., 1:208.
123. Foucault, *The Birth of the Clinic*, 140–41.

124. Rymer and Peckett Prest, *Varney the Vampyre; or: The Feast of Blood*, 1:203.
125. Ibid., 1:206.
126. 'HEREFORD, Sunday,' *The Times*, 1832, 7.
127. Ibid.
128. 'Apprehension of a Gang of Resurrectionists,' *The Times*, 1832, 4.
129. Ibid.
130. This word also positions the bodysnatchers, quite unusually, as victims: all social classes despised bodysnatchers, and yet, they become sacrificial victims if compared to the enraged crowd.
131. Charles Dickens, *Oliver Twist*, 1994th ed. (Oxford: Oxford University Press, n.d.), 74. Dickens's emphasis.
132. Rymer and Peckett Prest, *Varney the Vampyre; or: The Feast of Blood*, 1:207.
133. Ibid., 1:207.
134. 'Report from the Select Committee for Anatomy,' 71.
135. Ibid.
136. Rymer and Peckett Prest, *Varney the Vampyre; or: The Feast of Blood*, 2:208.
137. Ibid., 2:807.
138. Ibid., 2:819.
139. Although, they would not have explained the brick found in its place. I could find no record of bodysnatchers replacing bodies with other objects.
140. This is also a recurrent theme, for instance, in the writing of Edgar Allan Poe. In particular, in J. Gerald Kennedy, 'Poe and Magazine Writing on Premature Burial,' *Studies in the American Renaissance*, 1977, 177 Kennedy notes the short story 'The Premature Burial', which presented a cataleptic man terrorized by the idea of being accidentally buried alive, was published on *The Philadelphia Dollar Newspaper* in 1844, therefore rather close to *Varney*.
141. Rymer and Peckett Prest, *Varney the Vampyre; or: The Feast of Blood*, 2:819.
142. Ibid., 2:818.
143. Ibid., 2:812. As the Dover edition is a reprint of the original penny blood, we encounter here one of the typical issues of cheap serialized fiction, that is, mistakes in page sequence. After page 819, the page numeration starts again from 812.
144. Ibid., 2:813.
145. Hurren, *Dying for Victorian Medicine*, 74.
146. Ibid., 75.
147. Ibid.

148. Ibid.; See also A.W. Bates, '"Indecent and Demoralising Representations": Public Anatomy Museums in Mid-Victorian England,' *Medical History* 52, no. 1 (2008): 17.
149. In Paul Vita, 'Returning the Look: Victorian Writers and the Paris Morgue,' *Nineteenth-Century Contexts* 25, no. 3 (September 2003): 242, Vita notes that at the time in which *Varney* was being issued, the British traveller would experience, and describe, the old morgue, situated in the Quai du Marché Neuf. There, a glass window separated the dead and the living, guaranteeing safe distance between the living observers and the observed corpses. The new Paris morgue, built in the Ile de la Cité after the old morgue was demolished in 1867, added 'curtains to conceal the changing of the scene', which reinforced the spectator's impression of watching a show. Ibid.
150. Ibid., 241.
151. Ibid., 242.
152. Ibid.
153. Richard Davenport-Hines, *Gothic: Four Hundred Years of Excess, Horror, Evil and Ruin* (New York: North Point, 1998), 248. Quoted in Rymer, *Varney the Vampyre; or: The Feast of Blood*, 17.

List of Works Cited

'Apprehension of a Gang of Resurrectionists.' *The Times*. April 19, 1832, 4.

Baldick, Chris. *In Frankenstein's Shadow—Myth Monstrosity, and Nineteenth-Century Writing*. Oxford: Clarendon Press, 1987.

Bates, A.W. '"Indecent and Demoralising Representations": Public Anatomy Museums in Mid-Victorian England.' *Medical History* 52, no. 1 (2008): 1–22. http://www.scopus.com/inward/record.url?eid=2-s2.0-47749146135&partnerID=tZOtx3y1.

Cooper, Bransby B. *The Life of Sir Astley Cooper, Bart, Vol. I*. London: John W. Parker, West Strand, 1843.

Dickens, Charles. *Oliver Twist*. Oxford: Oxford University Press, 1994.

Foucault, Michel. *The Birth of the Clinic*. London: Routledge, 2003.

Gilbert, Pamela K. 'Sensation Fiction and the Medical Context.' In *The Cambridge Companion to Sensation Fiction*, edited by Andrew Mangham, 182–95. Cambridge: Cambridge University Press, 2013.

Grainer, L.J. 'A Medical Student in Search of a Supper.' *The Penny Satirist*, no. 164 (1840): 3–4.

'HEREFORD, Sunday.' *The Times*. January 10, 1832, 7.

Hurren, Elizabeth T. *Dying for Victorian Medicine—English Anatomy and Its Trade in the Dead Poor, c. 1834–1929*. London: Palgrave Macmillan, 2012.

Kennedy, J. Gerald. 'Poe and Magazine Writing on Premature Burial.' *Studies in the American Renaissance*, 1977, 165–78.

Liggins, Emma. 'The Medical Gaze and the Female Corpse: Looking at Bodies in Mary Shelley's Frankenstein.' *Studies in the Novel* 32, no. 2 (2000): 129. http://proxygsu-ogl1.galileo.usg.edu/login?url=http://search.ebscohost.com/login.aspx?direct=true&db=fth&AN=3371644&site=eds-live%5Cnhttp://content.ebscohost.com/ContentServer.asp?T=P&P=AN&K=3371644&S=R&D=fth&EbscoContent=dGJyMMTo50SeprY4zOX0OLCmr0ueprZSsae.

'Report from the Select Committee for Anatomy.' London, 1828.

Richardson, Ruth. *Death, Dissection and the Destitute*. Chicago and London: The University of Chicago Press, 1987.

Romano, Terrie M. 'Gentlemanly Versus Scientific Ideals: John Burdson Sanderson, Medical Education, and the Failure of the Oxford School of Physiology.' *Bulletin of the History of Medicine* 71, no. 2 (1997): 224–48.

Rymer, James Malcolm. *Manuscripts from the Diary of a Physician, 2 Vols*. London: Lloyd, Edward, n.d.

———. *Varney the Vampyre; or: The Feast of Blood*. Edited by Curtis Herr. 2008th ed. Crestline, CA: Zittaw Press, n.d.

Rymer, James Malcolm, and Thomas Peckett Prest. *Varney the Vampyre; or: The Feast of Blood*. Edited by E.F. Bleiler. New York: Dover Publications, 1973.

Senf, Carol A. *The Vampire in 19th Century English Literature*. Bowling Green, OH: Bowling Green State University Popular Press, 1988.

Smith, Thomas Southwood. *The Use of the Dead to the Living*. London: Baldwin and Cradock, 1828.

Sparks, T. 'Surgical Injury and Narrative Cure in Wilkie Collins's Poor Miss Finch and Heart and Science.' *Journal of Narrative Theory* 32, no. 1 (2002): 1–31.

Sparks, Tabitha. *The Doctor in the Victorian Novel*. Farnham: Ashgate, 2009.

Twitchell, James B. *The Living Dead: A Study of the Vampire in Romantic Literature*. Durham: Duke University Press, 1981.

Vita, Paul. 'Returning the Look: Victorian Writers and the Paris Morgue.' *Nineteenth-Century Contexts* 25, no. 3 (September 2003): 241–55. https://doi.org/10.1080/0890549032000153876.

Wood, Jane. *Passion and Pathology in Victorian Fiction*. Oxford: Oxford University Press, 2001.

CHAPTER 4

Underground Truths: *Sweeney Todd*, Cannibalism, and Discourse Control

On the 21st of November 1846, the weekly issue of Lloyd's *People's Periodical and Family Library* featured the first instalment of a series innocently titled *The String of Pearls*. Under the title, a broad illustration showed a weeping girl sitting at a kitchen table, in the company of a gentleman. The gentleman's expression is concerned, and a little dog anxiously looks at the distressed girl. The domestic scene is carefully crafted to attract the reader's attention, hinting at an exciting story behind the girl's tears. Yet, nothing transpires from the illustration, or the title, about the lurid story of human flesh-pies better known to us as *Sweeney Todd, The Demon-Barber of Fleet Street*. The disappearance of the string of pearls from the title of subsequent rewritings and adaptations is unsurprising, as the jewel soon ceases to have a key role in the story, overcome by the striking presence of one of Victorian popular fiction's most formidable villains: the 'demon' barber Sweeney Todd. Similar to other penny blood villains, the barber is a murderer, a robber, and a cunning cheater; what singles him out, making his callousness transcend humanity and become demonic, is his role as the facilitator and chief supplier of a ghastly business partnership with his pie-maker neighbour, Mrs. Lovett. In this chapter, I explore how this partnership reflected the discourses and spaces of the Anatomy Act, and the imaginary—and not-so-imaginary—horrors of bodily disintegration in the Victorian metropolis.

A. Gasperini, *Nineteenth Century Popular Fiction, Medicine and Anatomy*, Palgrave Studies in Literature, Science and Medicine,
https://doi.org/10.1007/978-3-030-10916-5_4

The barber murders his customers, the ones who will not be immediately missed, such as merchants or sailors, and who happen to be in possession of sums of money or valuables. Todd drops them in his cellar through a mechanical chair mounted over a trapdoor, breaking their necks; if they survive, he 'polishes them off' with his razor. The bodies are then butchered and transformed into 'pork' and 'veal' steaks, which are stored in Lovett's cellar and there turned into meat pies by a cook unaware of the origins of the material. When the cook realizes he is a prisoner in the cellar, and perhaps starts suspecting where the 'meat' comes from, Lovett and Todd 'dismiss' him and get a new cook. The series relates the end of this partnership, following the murder of Mr. Thornhill, a sailor who, unlike Todd's previous victims, has friends who come looking for him. The search also involves beautiful Johanna Oakley—the weeping girl of the illustration—whose fiancée, Mark Ingestrie, should have returned from his travels at sea, and was the reason why Thornhill was on land at all. He was meant to give Johanna a token from Mark, the eponymous string of pearls, and to bring her the news that the young man was lost at sea. Johanna impersonates a young boy, Charles, to take service at the barber shop when the police and the sailor's friends start focusing their investigations on the barber. The place of barber assistant has been vacant since Tobias Ragg, Sweeney Todd's previous apprentice, was shut away in a mad-house after he started suspecting his employer of murder. Tobias will finally manage to escape his prison, as will the current cook at Lovett's, Jarvis Williams. Williams, starving and destitute, applied for a job at the pie-shop in Bell Yard, and his timing was perfect: Lovett needed to replace her cook, and Williams took his place in the basement. After a while, though, he pieces together the truth behind the pie-making business and plans a daredevil escape. He mounts on the platform that hauls up the pies into Lovett's shop by way of a windlass, hiding under the tray of freshly cooked pies. As he reaches the top, he jumps up and screams the terrible truth to the customers: they are gorging themselves on human flesh. Mrs. Lovett dies, not because she is unable to cope with the events, but because Todd had poisoned her a few hours earlier. Conscience was starting to take its toll on the pastry-cook, which prompted Todd to make sure that she never compromised his cover. Finally, Todd is hanged, and Johanna is reunited with Mark Ingestrie, who is revealed to be Jarvis Williams. The story closes on Lovett's last living customer, an old man who still needs a drop of brandy when he remembers how much he loved his 'veal' pies.

The countless rewritings of this story make it probably the only penny blood to be famous outside academic circles, and in scholarly circles, *Sweeney Todd* is still an object of analysis and debate. Its authorship, for instance, is still controversial: traditionally, the text was attributed to Thomas Peckett Prest and, while Helen Smith has produced convincing evidence in favour of Rymer, other scholars remain sceptical.[1] Crone takes yet a different stand, arguing that any debate around the authorship of penny bloods is pointless, unless it is aimed at highlighting the genre's fundamental homogeneity.[2] I do not completely agree with her point: as I have discussed in previous chapters, casting light on penny blood authors may open new perspectives for analysis of the narratives. However, it is not my purpose here to add to the authorship debate, but rather to analyse the role of this highly successful penny blood as a vehicle of discourses connected with the world of medicine and dissection.

In Chapter 1, I have explained how there is a consensus among literary scholars that *Sweeney Todd* was deeply rooted in the socio-historical context of the mid-nineteenth-century, and that it elaborated anxieties specific to the lower section of the social spectrum, which Lloyd's productions explicitly addressed. Significantly, scholarship on this narrative tends to highlight the connection between the theme of cannibalism and working-class concerns about physical integrity after death in the Anatomy Act era.[3] Indeed, while the medical discourse is more elusive in this than in the other penny bloods examined in this book as there are no doctors amongst the characters, this very elusiveness is crucial to chart the medical discourse in *Sweeney Todd*, a story in which the impossibility of speaking about certain topics is the key to understanding the power dynamics between characters. London was already familiar with popular myths of butchery and cannibalism before the narrative was serialized.[4] Still, the presence of a barber cutting corpses into pieces in an underground space is meaningful in a historical context of underground dissection rooms and anxieties about the butcher-like procedures that characterized medical education and practice. Popular conscience likened the work of surgeons and anatomists to butchery, and the burkers' incidents literalized the concept of retailing the human body as if it were butcher's meat. The combination of these elements triggered a set of anxieties about cannibalism, the idea of cooking and consuming human flesh, related to the anatomy world. Furthermore, as early as 1948, Turner noticed the 'grim *double-entendre*' of the *Sweeney Todd* plot[5]; this double-entendre, which characterizes particularly the speech of the murderous couple Todd-Lovett, can be examined against the background of the obscure and complicated language of the Act. Finally, the murderous couple is a model of industrial, indeed Utilitarian, efficiency that resembles the way in which the Anatomy Act put the powerless members of society in a position comparable to that of a portion of meat for a grinder, while benefiting chiefly, if not entirely, the more powerful echelons of society.

The use of cannibalism as a metaphor for heartless treatment of the pauper was already part of British reading culture: in 1729, Jonathan Swift's satirical pamphlet *A Modest Proposal* suggested that the Irish pauper should start, not only selling their children as choice meat to the rich, but that the population should start breeding them with precisely that purpose. 'A young healthy child well nursed is at a year old a most delicious, nourishing, and wholesome food' he argues,[6] and the system would reduce poverty, promote marriage, motherly love, and make husbands more loving of their pregnant wives.[7] *The String of Pearls* plays on a similar representation of a brutalized, bestial society where people who will not be missed become cattle for the butcher's knife to solve the problem of the rising number of the destitute. Todd and Lovett's business, like the Anatomy Act, was a perfect solution: they both ensured that nothing was wasted, minimized the costs while maximizing the income, and were, in their efficiency, perfectly soulless.

In this chapter, I suggest that *Sweeney Todd* reiterated anxieties about the underground space in relation to anatomical practice in the metropolis, especially if we consider that the Anatomy Act did not solve the intrinsic unfairness of the body trade. This matter, as Powell and Crone point out, was decidedly relevant to the readership of the narrative. Simultaneously, the narrative proposed an alternative, cathartic solution to the unfair system: instead of the secrecy and obscurity that characterized the language of the Act,[8] and the proceedings of the medical fraternity, the truth is seen, uttered, and believed, and the system that concealed the truth is dismantled.

1 A Monstrous Partnership: Burking, Dissecting, and Pie-Making

Todd and Lovett, the managers of the narrative's monstrous production system, seldom appear together in the original series, which, unlike later adaptations, did not suggest in any way a romantic connection between them. Yet, they are undeniably a couple, *the* couple: their business relationship propels the action in the plot, and they were the chief medium the original narrative used to convey the double-entendre Turner noted. While later adaptations, particularly Bond's theatrical adaptation,[9] tended to humanize Todd and Lovett, turning the slippage of meaning in their speech almost into a joke between the murderous duo and the audience in the theatre, the original 1840s penny blood was an altogether different matter. The element that emerges most forcefully throughout the whole narrative is that there is nothing human in Sweeney Todd and Mrs. Lovett with which the reader can empathize. Whereas the intrinsic humanity of the theatrical Todd and Lovett creates a guiltless complicity between spectator and characters, the original narrative leaves for the reader clues to the truth that produce an uneasy, unwelcome proximity to the couple's unspeakable crimes.

As in *Manuscripts* and *Varney*, the concept of monstrosity in *Sweeney Todd* implies lack of humanity, departure from nature. As with *Varney*, the heroes in the story play second fiddle to the monstrous villains, who are the undisputed centre of the narrative. Unlike Varney, however, Sweeney Todd has no redeeming qualities to speak of, and Mrs. Lovett, though in part a victim of the demon barber herself, does not awaken the reader's sympathies. This repulsion originates in the fact that the couple commit several unpardonable sins at once: they are serial killers who also involve other people in an act of cannibalism, which simultaneously contaminates the community and wipes away their victims' identity. It is therefore unsurprising that Sweeney Todd later acquired the sobriquet of 'demon barber': the couple is repeatedly characterized as diabolical and, although neither of them is an actual supernatural monster, they display several physical and behavioural traits typical of preternatural figures.

The conspicuous eeriness of Todd and Lovett's physical aspect is the first clue the reader receives to solve the mystery of the narrative. The description of Sweeney Todd is not flattering: he is 'a long, low-jointed, ill-put-together sort of fellow, with an immense mouth' and his 'huge hands and feet' make him 'quite a natural curiosity'.[10] The narrator ironically notes that, considering his profession, the barber's most extraordinary feature is his hair, which resembles 'a thickset hedge, in which a quantity of small wire had got entangled'.[11] This description purposefully frames the barber's body as disproportionate, a deviation from nature.[12] The adjective 'ill-put-together' suggests an artificial breach of the natural composition of the human body, as if Todd has been assembled, rather than born. Moreover, the emphasis on the disproportionate size of Todd's frame gives his figure an ogreish quality, particularly the 'immense mouth', which ominously suggests the need for commensurate meals. His features seem planned to trigger the idea of the monstrous in the mind of the reader, connecting the barber's body to that of the widely popular figure of Frankenstein's Monster, a connection that is even more evident than the one found in *Varney*. In *Frankenstein*, descriptions of the Monster emphasize his 'gigantic stature' and disproportionate frame.[13] Frankenstein's Creature's aspect is deformed and 'more hideous than belongs to humanity',[14] which automatically identifies him as an alien, and Todd's 'ill-put-together' body reiterated this characterization. It is worth nothing that the Monster is intrinsically linked to the world of anatomy and body traffic, being the result of the assemblage of parts from different fresh bodies stolen from cemeteries, and this trait of the Monster's bodily history resonates in Todd's monstrous physicality. The barber's awkward body looks as if it has been inexpertly pieced together, and his daily activity consists in dismembering human bodies in a cellar with the purpose of destroying them completely. The idea of 'ill-putting-together', and the apparently inevitably destructive tendency of the inaptly constructed subject seem to belong to the Creature's and Todd's body alike.

The unnatural quality of Sweeney Todd's physicality emerges also in his voice and eyes. Todd's peculiarly un-natural laugh is 'short', 'disagreeable', 'unmirthful', and 'sudden', possibly triggered—the narrator suggests—by the memory of 'some very strange and out-of-the-way joke'.[15] The narrator compares it to the bark of the hyena, and claims it left the listener under the impression that it could not have come 'from mortal lips', so that they looked 'up to the ceiling, and on the floor, and all around them'.[16] While what they expect to see is not specified, it is assumed to be something supernatural and malignant, and thus the inarticulate, but spontaneous sound of his laugh gives a glimpse of the barber's inhuman nature: Todd is at his most natural when he sounds most unnatural.

Todd's physical description closes on the observation that 'Mr. Todd squinted a little to add to his charms'.[17] Victorian readers of popular fiction would be used to a reference to the eyes in characters' descriptions. Some of the most famous Dickensian villains' eyes match their nature, like Daniel Quilp's 'restless, sly and cunning' eyes,[18] or Artful Dodger's 'little, sharp, ugly eyes'.[19] Penny blood authors adopted the same strategy and, usually, something odd in the gaze gave away the villains, in the same way the eyes of the heroes and heroines mirrored their goodness. Johanna's eyes, for instance, are 'of a deep and heavenly blue'.[20] Sweeney Todd's eyes 'squint'. Primarily, this means that he is affected with strabismus; secondly, it suggests that he does not look directly at people or things. Hence, his eyes are simultaneously deformed and impossible to decipher, they can look without returning the gaze. They are, in brief, 'simply wrong'.[21]

Mrs. Lovett's wrongness also surfaces in her body and eyes. At first sight, the pastry cook is as sensual and charming as her pies. The inviting look and delicious taste of the pies and Mrs. Lovett's beauty are one thing, because, muses the narrator, 'what but a female hand, and that female buxom, young and good-looking, could have ventured upon the production of those pies [?]'.[22] Mrs. Lovett's body is sensual and, although it is not explicitly stated, her customers imagine that by eating her pies, they are partaking of that sensuality. The pies themselves are described as peculiarly sensual, meaning that they gratify the senses, primarily as culinary delicacies, but also and more subtly as an extension of Mrs. Lovett's sensuality. The 'construction of their paste' is 'delicate'; the 'small portions of meat' they contain are 'tender'; they are 'impregnated' with the delicious 'aroma' of their gravy; the fat and meagre meat are 'so artistically mixed up' that eating one of Lovett's pies is a 'provocative' to eat another.[23] This description is constructed so as to be positively mouth-watering; yet, most of the adjectives, if taken out of context, are applicable to female beauty, as smallness, tenderness, delicateness, and proportionate appearance are highly appreciable qualities in the Victorian female body. Moreover, the 'impregnated aroma' and the 'provocative' trait of the pies would not be out of place in a boudoir scene. Lovett's pies are manufactured to be as captivating as is their cook: all of Mrs. Lovett's young customers, the clerks and law students from the Temple and Lincoln's-inn, are 'enamoured' of her, and they toy with the thought that Mrs. Lovett made the pie they 'devoured' especially for them.[24] The implicit suggestion is that they are actually fantasizing about devouring the pastry-cook herself, in the more unchaste meaning of the word.[25]

This wantonness, though, is soon framed as something eerie, as the narrator explains that Lovett exploited her admirers' appreciation to induce them to buy more pies, smiling more often at her best customers.

This game was 'provoking to all except to Mrs Lovett', while the 'excitement' (yet another ambiguous word in an ambiguous context) it generated 'paid extraordinarily well', inducing the most exuberant customer to consume pies 'until they were almost ready to burst'.[26] At this point, the narrator adds a darker layer to the picture, remarking that other customers, who were only interested in the pies, judged Lovett's smile to be 'cold and uncomfortable—that it was upon her lips, but had no place in her heart—that it was the set smile of a ballet-dancer, which is about one of the most unmirthful things in existence'.[27] Others still, while conceding the pies were excellent, 'swore that Mrs Lovett had quite a sinister aspect, and that they could see what a merely superficial affair her blandishments were, and that there was "a lurking devil in her eye"'.[28]

The comparison of Mrs. Lovett to a ballet dancer could be extended to her whole physicality. The beautiful pastry-cook is performing a dance for her customers, made of ritualized, rehearsed movements, each one devoted to selling more pies. As the description of Mrs. Lovett grows darker, the concept of artificiality, of something 'ill-put-together' that resembles humanity but fails to fully succeed, surfaces in the body of Todd's business partner. The eyes are the only place where something of Lovett's true nature can be guessed, and what they show is peculiarly un-natural. Mack notes that, besides being reminding of such works as Byron's *Mazeppa*, the phrase 'a lurking devil in her eye' was typical of character description in Gothic fiction.[29] Therefore, Lovett's sensual and amiable façade disguises an evil soul.

I would add a further layer of analysis to the concept of 'evil' in the characters of Todd and Lovett by examining their connection with the supernatural. The two are no vampires, and yet, the narrative hints at something preternatural about them, which, if it does not correspond to their actual nature (in the end, they are both human), is definitely something the two characters very closely resemble. Mrs. Lovett's behaviour and some of the adjectives used to describe her connect her to the figure of the witch. With her smiles, she charms her customers into eating more pies, keeping control over them and her invoices simultaneously. One of her customers even calls her 'charmer',[30] which is meant as a compliment on her beauty, but also defines her effect on people. She casts her spell by exploiting her victims' lust, using her sex-appeal to encourage her customers to eat more, giving the process of eating a sensual connotation. The malignity of the spell is announced: Lovett's customers gorge themselves on the pies until 'they are almost ready to burst', as if the meat in the pies has preserved its cadaverous chemistry and emanates explosive gases.[31] As the victims of a spell in a fairy tale, they end up ruining themselves through the unchecked, sensual consumption of food that goes on in Lovett's premises. Since both the food and the cook are sensual, the sickness that ensues is doubly shameful: the customers yield to both gluttony and lust. The image of the 'devil' lurking in Mrs. Lovett's eyes seals her characterization as a witch, a dangerous and essentially monstrous character whose enchantment manages to deceive even those customers who, although claiming to be immune to her charming looks, are not immune to the charm of the pies, and become unwitting participants in Lovett's ghastly cannibal istic banquet.

As for Sweeney Todd, the barber becomes increasingly vicious as the narrative progresses, until he is explicitly likened to the devil. In a moment of malicious happiness, Todd resembles 'some fiend in human shape, who had just completed the destruction of a human soul'.[32] The use of the word 'fiend' in this passage is meaningful. This ancient term basically means 'enemy', which connotation also relates to the world of supernatural forces and magic, acquiring the definition of 'demon or evil spirit, the devil itself as the enemy of mankind, and, finally, a person of supernatural wickedness'.[33] The image of the 'destruction of the human soul' reinforces Todd's connection with the demonic in Christian sense. Not only does he perform mischief, but he enjoys it, as a devil would. Furthermore, Todd's characterization as enemy with the meaning of 'devil' becomes explicit after the reader has been given enough clues to suspect him of murder.[34] He becomes 'the arch-enemy of all mankind' in the eyes of Tobias, behind whose shoulders he stands, unseen, making 'no inept representation of the Mephistopheles of the German drama'.[35] Even his witch-like business partner, towards the end of their relationship, identifies him as the destroyer of her soul, exclaiming bitterly: 'Oh. Todd, what an enemy you have been to me!'[36]

Todd and Lovett's characterization, therefore, includes elements of the devil and the witch, two interrelated figures of Christian folklore. This adds a further degree of monstrosity to their partnership, as if to mark the peculiar viciousness that comes with being a *commercial* association based on murder: Todd and Lovett are inhuman because they are disconnected enough to commit multiple murders *and* to recycle their victims as food. The couple's inhumanity emerges in all its devilishness as soon as it becomes clear that Lovett's pies are filled with the flesh of Todd's victims. This moment coincides with the scene in which the local tobacconist's wife, Mrs. Wrankley, asks Lovett's permission to put up in her pie-shop a bill asking for information on the disappearance of her husband. The man has been killed by Todd, who listens 'impenetrably' as Lovett reads the bill.[37] Then, the barber comforts the woman and suggests that she buy a pie and eat it, 'lift[ing] off the top crust', declaring that she would 'soon see *something* of Mr. Wrankley'.[38] Although the widow (for that is what the narrator calls her, although she has not yet been notified of Mr. Wrankley's death) recoils from Todd's 'hideous face', she accepts the pie because it is 'very tempting', and Todd's words even raise her hopes.[39] The scene shows the full extent of the barber's monstrosity: by playing this macabre prank on the widow, of which only he, Mrs. Lovett, and the reader can be aware, Todd enjoys raising Mrs. Wrankley's hopes as he feeds her, quite possibly, her own husband. Mrs. Lovett, who but a few moments earlier protested that she hated her business partner, does not refrain from selling the pie to the widow. Powell argues that, although Lovett's active involvement in the actual killing remains uncertain, she 'knowingly and ruthlessly' sells the final product, which diminishes her womanliness.[40] Lovett's lack of womanly qualities such as love, tenderness, and compassion emerges clearly in her involvement in the cruel joke Todd makes at Mrs. Wrankley's expense, which emphasizes her monstrosity.

Todd and Lovett's characterization as a murderous *couple*, a commercial partnership, can be related to the commercial partnerships formed by the Edinburgh and London burkers, which was also devoted to the commodification of dead bodies and had attracted the attention of the press between 1829 and 1832. In both cases, the men worked in couples, and had female partners whose degree of involvement in the murders remained uncertain. The news coverage of burkers' cases was massive, occurring almost daily in the month in which each case broke, which contributed to making burkers and bodysnatchers a substantial, and sensational, part of the life of the British public. The 'Italian Boy' case, particularly, was sometimes the subject of two, or even three articles in the same issue of a newspaper,[41] besides inspiring ballads[42] and even a 'genuine edition' of the trial by Pierce Egan.[43] Such was the resonance of the cases that, about a decade later, in 1841, Burke's trial for the Edinburgh murders occupied nineteenth pages of *The Chronicles of Crime, or: The New Newgate Calendar* by Camden Pelham (a pseudonym), 'of the Inner Temple, Barrister-at-Law'. Pelham's account of the murders and trial, a mixture of almost-accurate facts and plain inaccuracies, as it was usual with the Newgate Calendar genre, focuses on Burke, as he was the only one to undergo death sentence, but he never spares Hare his disdain and represents him as cunningly coward in his betraying Burke by turning king's evidence.

In the months that followed the discovery of the London homicides, newspaper reports of corpse-stealing or attempted burking abounded, tapping into the public's outrage and fear. In December 1831, right after the London burkers were discovered, the *Times* published an article about a spectacular cadaver theft in Dublin. A whole gang of resurrectionists allegedly broke into a first-floor apartment and stole the body of an old woman right in front of her mourners. The article claimed that they made it downstairs before any of those present could stop them, and disappeared into the night, shamefully dragging the corpse by the shroud on the mud of the street.[44] Richardson connects the daredevil quality of this theft to the increase in prices paid for corpses, which made the resurrectionists more daring[45]; however, the way the article itself is constructed is significant. The detailed description of each trait that might contribute to portray the bodysnatchers as sacrilegious and disrespectful, such as the 'revoltingly indecent' element of the body being dragged in the mud by the shroud, suggests that the piece aimed to stir up animosity towards resurrectionists.[46]

The press also contributed to spreading the idea that burking was practised by 'clapping' pitch plaster over the mouth and nose of the victim to suffocate them, the so-called pitch plaster myth. Sometimes, pitch plaster aggressions were made in jest.[47] In other instances, victims, especially children, reported having been attacked, usually by one or two men who placed plaster over their faces, as in the case of young Charles White, in November 1831.[48] Notably, White stated that one of his assailants wore a smock-frock; analysing the article, Wise observes that the smock-frock was the detail of Bishop and Williams's outfits on which the newspapers focused their attention, creating a connection between these garments and burking in the popular mind.[49] I would add that this attests to the pervasiveness of the press campaign against bodysnatching and burking, which portrayed the people engaged in the body traffic as demons disguised as common people. Crone notes that penny bloods tended to provoke a *frisson* in the reader by making their villains familiar, everyday figures,[50] as is the case of Sweeney Todd, the barber. Although illustrations tend to represent Todd wearing an apron, rather than a smock-frock, *Sweeney Todd* catered to the idea, consolidated in the public's mind by the news coverage of the burkers' cases, that a monster bent upon making money out of murdered bodies could have the outward appearances of a common worker.

The female presence in the burkers' cases added to the repulsion they generated. In January 1829, the *Times* contemptuously described Helen M'Dougal, Burke's common-law wife, as 'utterly destitute of shame and common prudence, as she [was] of humanity', and defined her relationship with Burke a 'hideous sympathy'.[51] The article also emphasized the Hares' status of married couple and the reporter did not disguise his certainty that both husband and wife were guilty. The press, therefore, represented both Helen M'Dougal and Margaret Hare as the accomplices of their partner's crimes. Even Pelham's historically doubtful account of the trial in the *Chronicles of Crime* echoes the harsh judgement the press gave of the Edinburgh burkers' wives more than a decade earlier: he describes Helen M'Dougal as a 'polluting presence'[52] and, as regards Mrs. Hare, he focuses on the fact she went to give her evidence holding a baby in her arms, a detail that had attracted attention at the time. Whatever reasons had Margaret Hare's for doing this, it did not help her case. Mr. Cockburn, part of the counsel for Helen M'Dougal, declared to the court that her face displayed 'every evil passion' and, as for the 'miserable child in her arms', rather than receiving 'a look of maternal tenderness in his distress' was treated with 'harshness and brutality'.[53] He declared that Mrs. Hare, in fact, 'eye[d] it in such a way as added to her malign aspect'.[54] Lutenor's portrait of Mrs. Hare echoes this same impression: Margaret Hare's brow is knitted, the lips thin and tense, her expression halfway between malevolent defensiveness and cunning that is not made to be attractive. The infant in her arms has a long, serious face and sits stiffly in his mother's arms, with every impression of being uncomfortable.[55] These previous portraits emerge in Pelham's account of Mrs. Hare in the witness box: he frames her as a peculiarly unloving mother, one who looks at her child 'ill of hooping-cough' (sic.) with 'aversion and hatred, shaking it and squeezing it [...] with the expression of a fury in her countenance', while 'even the tigress' would show 'maternal tenderness [...] for her whelps'.[56]

As for the London burkers' wives, Sarah Bishop and Rhoda Williams, there is comparatively little newspaper material on them,[57] and what there is shows that they were relatively cooperative. When they appeared before the magistrate on November 11, 1831, both women released statements, although they were informed that these could be used to incriminate them.[58] When Rhoda was apprehended as an accessory to the murder of Fanny Pigburn, she wept 'bitterly' upon hearing the charges and, though again she was informed that her statements could be used against her, she fully cooperated.[59] The press represented Sarah and Rhoda as quite docile, even scared, and did not ascribe to them the negative moral qualities that characterized the portrait of Helen M'Dougal and Margaret Hare. The public's judgement, however, was harsher. After the trial, Sarah and Rhoda moved to the neighbourhood of Paradise Road; this was, Wise points out, particularly unattractive, and yet the arrival of 'the kin of burkers' was cause of deep concern, and the newspapers reported that mothers would forbid their children to play outside as long as Sarah and Rhoda resided there.[60]

The female presence in the burkers' cases might have contributed to the creation of the image of the criminal couple in the public's mind, which made it easier for the audience to accept Todd and Lovett's murderous partnership. Mrs. Lovett is a peculiarly un-loving, unwomanly woman, a trait that characterized also the burkers' companions, especially in the Edinburgh trial, and becomes melodramatically exaggerated in the fictional character. Cold, calculating Mrs. Lovett cannot be a wife, nor can she be a paramour: her relationship with the devilish Todd is pure business.

The idea of business partnership is key to the connection between the burkers' cases and the narrative of *Sweeney Todd.* The element that singled out burking was that, for the first time, homicide was committed as a commercial transaction. It put a price on the human *body*, equalling the individual to livestock. John Adolphus, speaking for the prosecution during the London burkers' trial, stated that '[n]othing but the sordid and base desire to possess themselves of a dead body in order to sell it for dissection had induced the prisoners [...] to commit the crime for which they were now about to answer'.[61] Similarly, Rosner notes that, while murder in Edinburgh was a relatively uncommon occurrence, and mostly passion-related, the Burke and Hare murder generated panic because they were conspicuously *not* driven by passion, and suggested that there was 'literally a price [...] upon every head'.[62] Burking practically performed that 'commodification' of the human body that, Powell convincingly argues, industrial economy performed metaphorically on the body of the workers.[63]

The public was not easily distracted from the fact that the medical community was at the other end of the commercial transaction. The fact that Robert Knox was never tried for the Edinburgh murders was bitterly regretted, and indignation was voiced in several quarters. Sir Walter Scott, for instance, declined the requests of the surgeon's friends to be part of the committee appointed to prove Knox's innocence. Scott declared that he would not help to 'whitewash this much to be suspected individual'.[64] The newspapers too, though not mentioning him directly, expressed outrage at the fact that Knox did not stand trial, thus increasing suspicions in the public's mind regarding the role of medicine in the murders. The *Times* emphatically proclaimed: 'what tales still remain untold! Bodies, never interred, have been purchased without question or scruple. Is this also to pass without further investigation?'.[65] Likewise, the title of the *Times* article reporting the trial, 'The late Horrible Murders in Edinburgh, to Obtain Subjects for Dissection',[66] reminded the public that burking and anatomy were directly connected. Two years later, the London burkers' incident linked medicine and murder again. The medical community's position with respect to the homicides was not as ambiguous in this case—in fact, as soon as he suspected the corpse he was being offered could be a murder victim, Richard Partridge went for the police. Yet, the case rekindled the debate around the shortage of subjects for anatomy courses and revived the fear that criminal individuals may resort to killing to provide the commodity. From the press to cheap serialized fiction, the step was short. In 1841, the *Newgate Calendar* published the first edition of *Murderers of the Close*, the account of Burke and Hare's crimes in Edinburgh. In 1846, the first episode of *The String of Pearls*, a story about a murderous couple cutting homicide victims into pieces to sell them, was issued.

From this perspective, Sweeney Todd's profession provides further ground for analysis, as it adds a further nuance to his participation in the process of butchering the 'meat'. When the first issue of *Sweeney Todd* was published, barely a century had passed since the Company of Barber-Surgeons split in 1745. Mack, however, notes that in the popular mind, the connection was not so easily untied: before the two professions became separate, in addition to the occasional razor cut, barbers shed their customers' blood also by performing minor surgery, tooth-drawing, sometimes amputations, and, in the case of Italian barbers, even dissections.[67] Such degree of intersection between the two professions had indeed long-lasting effects on nineteenth-century popular culture, which likened the surgeon to the butcher. The cartoon 'A Few Illustrations for Mr Warberton's [sic] Bill' (Fig. 1), by William Heath, explicitly connects

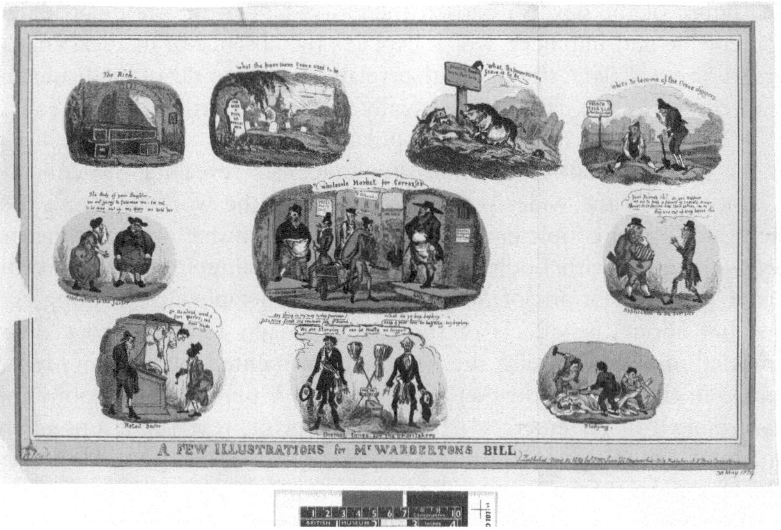

Fig. 1 Heath, William ('Paul Pry'). 'A few illustrations for Mr Warbertons Bill'. Print 1829.

the two trades. The print pictures a dystopic future under the Anatomy Bill, in which the jail, the workhouse, the hospital and the King's Bench have become retailers of human bodies that display price-per-weight placards in the same style as butcher shops. A doctor's servant purchases for his master pieces of human meat hanging from butcher's hooks, while the lower right-hand corner vignette, titled 'Studying', shows medical students savagely hacking a corpse with hatchet, hammer and saw.

Sweeney Todd, being simultaneously a barber and a butcher of human flesh, summarizes the popular representations of the surgeon. His detachment doubles as an extreme interpretation of the inhumanity that surgeons and anatomist s considered necessary to perform their tasks. The demonic barber Sweeney is as dystopic a figure as the medical students in Heath's cartoon, one that embodies the frightful possibilities of the commodification of the human body for dissection purposes. A further element that reinforces this trait in Sweeney Todd are the 'heads and bones' of his victims, which the police finds in the catacombs below St. Dunstan's.[68] These resemble the 'disintegrating bone, brain, trunk and decomposing flesh' that were left of a human body after dissection.[69]

The figure of the devilish barber, therefore, tapped into popular images of medicine and butchery, which explains the absence of medical characters in *Sweeney Todd*: the figure that relates to the world of medicine and its discourses in the narrative is actually the demon-barber himself.

If Todd's figure tapped into popular images that satirized the figure of the surgeon-anatomist, while simultaneously revealing the concerns it generated in the wider public (especially in the working class), then the Todd-Lovett couple embodied the mechanism that provided the surgeon-anatomist with bodies for dissection. Starting from Powell's and Crone's interpretations of *Sweeney Todd* as a metaphor of working-class anxieties about the effects of the Anatomy Act, I would move on to consider more specifically the system implemented by the murderous couple in the light of the Act itself, and the Utilitarian philosophy that underpinned it. As emerges from Heath's print, representations of the Anatomy Act as a cannibalistic system pre-dated its being voted in, were widely popular, and surface in the processes of killing, butchery, and cannibalism around which the plot of *Sweeney Todd* revolves.

Todd and Lovett's is a lucid and terrifying scheme in which both partners are committed to personal gain. Todd provides the commodity and rewards himself with his victims' possessions, which he hoards in his house. Lovett applies herself to increasing the sales of the final product to make the most from the commodity, while simultaneously achieving the crucial purpose of eliminating the victims' bodies. The system of feeding them to hungry customers tallied perfectly with London underworld's practicality, according to which the cows of London dairies were fed 'spent mash from the breweries' and 'market sweepings',[70] and animals 'awaiting slaughter', a pamphlet asserted, were fed 'cag-mag', a mixture of rotten meat and meat of diseased or otherwise second choice animals.[71] It is possible to detect in these dynamics, in the determination not to waste anything and devote everything to a useful purpose, an element of Utilitarian philosophy. As we have seen, that same Utilitarian philosophy eventually turned its attention to the problem of body supply for anatomy schools, individuating the perfect source of subjects in those human bodies that were perceived to represent a cost to the community. The Anatomy Act was an expression of this philosophy, indeed it was a masterpiece of Utilitarian thought. If not explicitly connected with the Anatomy Act in the narrative, Todd and Lovett's ultimate recycling enterprise, which disposed of 'unknown' people who would not be missed and turned them into food, concretized the anxieties expressed in Heath's print.

In the narrative, the pies produced through the monstrous recycling system are consumed by a whole community that stubbornly refuses to acknowledge that something is amiss in the neighbourhood, even when a suspicious smell of decay starts rising from below the pavement of their own church. Their indifference in effect endorses the killing and butchering that go on in the subterranean space of Todd and Lovett's shops, echoing the indifference to the Anatomy Act's impact on the pauper that characterized the spirit of the legislation, and the lack of effective opposition to its passing. The key element underpinning community endorsement in *Sweeney Todd* is the hiding and tabooing of the truth. The transmission of information is carefully policed by the two murderous partners, whose business stands on a solid basis of doctored information and silence.

2 Truth, Taboos, and Dénouement: Discourse Control and Power

It could be stated that *Sweeney Todd* is essentially a narrative about truth: as the action unfolds, truth is concealed, ignored, and discovered, while characters and readers are made privy to different bits of truth. Most of all, truth in *Sweeney Todd* is unspoken: insofar as it is not stated out loud, it is invisible and intangible, even though palpably *there*. As Turner's observation on the double-entendre characterizing the story suggests, language plays a crucial role in this dynamic. Todd and Lovett manage to hide the truth about their partnership for a long time through careful control of language, both their own and that of people around them. Notwithstanding their precautions, though, the whole narrative leads, unavoidably, to the final dénouement, when truth is finally spoken.

Todd and Lovett assert control firstly by preventing their interlocutors from understanding the real meaning of their statements. As with characters in *Varney*'s Bannerworth saga, Todd and Lovett's interlocutors can only suspect that the two are withholding information, but lack sufficient elements to ascertain this. Todd's speech is as maliciously ambiguous as Varney's, although the barber lacks all the charm the vampire baronet possesses. This ambiguity surfaces in Todd's words for the first time when the captain of the ship on which Mr. Thornhill served comes to Todd's barber shop to inquire about the missing sailor. When asked if he has ever seen the gentleman in question, Todd answers: 'Oh! to be sure, he came here, and I shaved him and polished him off', at which the two exclaim: 'What do you mean by polishing him off?' But Todd innocently replies: 'Brushing him up a bit, making him tidy'.[72] Like Varney, Todd is aware of his power over his victim's friends, and enjoys exercising it. The double-entendre that characterizes the barber's speech emerges in all its maliciousness in this exchange: as the reader well knows, the statement 'I polished him off' means that Todd has killed Thornhill. However, the barber easily modifies the sense of his words by deploying a slippage of meaning, in the same way Varney does to provoke Henry Bannerworth about Flora. As he speaks, Todd modifies the rules of communication based on his exclusive knowledge of the truth, of which he selects bits and pieces that are deprived of a finite meaning, disorienting the other characters and keeping control of the situation.

Mrs. Lovett uses a similar technique, as she blurs the meaning of her sentences to such an extent that she gives the impression of speaking in riddles. When Jarvis Williams, alias Mark Ingestrie, is hired as the new cook at Lovett's, she tells him that the old cook 'has gone to see some of his very oldest friends, who will be quite glad to see him'.[73] She adds that, should he accept the position, he must 'live entirely upon the pies' and 'agree never to leave the bake house', unless it is 'for good'.[74] She also assures Mark that she 'never think[s] of keeping anybody many hours after they begin to feel uncomfortable' and that 'everybody who relinquishes the situation, goes to his old friends, whom he has not seen in many years, perhaps'.[75] She is telling the truth, in a way: there is only one way to leave Lovett's basement, and that is death, which comes shortly after the moment in which a cook understands the truth. Therefore, the 'old friends' Mrs. Lovett alludes to are the ones that await the cook in the hereafter. However, she carefully phrases her information so that what appears is that the former employees leave to go back to their families and friends.

This dynamic bears remarkable similarity to how the relationship between the Act, the institutions, and the poor worked. Mark says that he is entering Lovett's basement out of 'poverty and destitution'[76]; his situation is as desperate as were the circumstances that compelled the poor to apply to the workhouse, which, after the passing of the Anatomy Act, made them candidates for the anatomist's slab. Moreover, Lovett's enunciation of the contractual clauses resembles the Act notices that were hung in the workhouses: while they summarized the Act's prescriptions, they were not explicit as to their meaning, which prevented the poor from understanding the full extent of the contract to which they were agreeing. Mark, the 'unknown poor', is being locked up and his name put down as a candidate for Lovett's meat grinder without him realizing it, because while she conveys the terms of the contract she manipulates the language, and Mark is not aware of this. Somehow, however, he perceives that she is not telling the whole truth. Noticing Lovett's cryptic phrasing, he wonders: 'What a strange manner of talking she has! [...] There seems to be some singular and hidden meaning in every word she utters'.[77] Being almost starved, Mark ignores his impressions, but his comment is a red flag for the reader: as with the reader of *Varney*, *Sweeney Todd*'s reader gradually receives enough clues to guess the truth, which puts them in a far worse position than that of *Varney*'s reader. They become the unwilling accomplices of the demon-barber, as they suspect the truth but are prevented from revealing it. Eventually, Mark understands the truth, but the pastry-cook and the barber have their methods to prevent the spreading of knowledge.

To control the diffusion of truth, Todd and Lovett create taboos for the other characters. This expedient can be explained with the Foucauldian concept of 'procedure of exclusion', that is, a strategy the person or persons who control discourse deploy to exclude other parties from power.[78] Foucault maintains that we are aware of the taboos placed on certain topics and argues that '[i]t does not matter that discourse appear to be of little account, because the prohibitions that surround it very soon reveal its link with desire and with power'.[79] Conforming to this principle, the first thing Todd does as soon as Tobias enters his employment, indeed his very first action in the series, is to dictate that the boy is not to speak a word about 'anything [he] may see, or hear, or fancy [he] see[s] or hear[s]' in the shop, or he will 'cut [his] throat from ear to ear'. To this, Tobias replies: 'Yes, sir, I won't say nothing. I wish, sir, as I may be made into veal pies at Lovett's in Bell-yard if I as much as says a word'.[80] Unsurprisingly, the barber's 'huge mouth' drops open, as he 'look[s] at the boy for a minute or two in silence, as if he fully intend[s] swallowing him', giving the reader a first glimpse of the truth.[81] The attempt to create a taboo for Tobias sees the boy accidentally coming dangerously close to the truth. The scene, particularly the detail of Todd's 'huge' open mouth and the suggestion that he intends to eat Tobias are the very first pieces that build the narrative's underlying discourse on cannibalism, the unspoken truth the barber strives to conceal. It is also the first clue the readers receive, which allows them to start immediately piecing together the truth. Tobias's simple remark casts a new, sinister light on the already eerie figure of the barber, creating an uncanny connection between the pies, the barber, and cannibalistic eating as if the spoken word, by virtue of some creative power of its own, could make truth real for the listener.

Of course, the barber cannot let this happen. From this moment onwards, Todd makes sure to prevent his apprentice from uttering the truth again, no matter how unwittingly. Firstly, he reduces Tobias to silence with physical violence. Then, he blackmails him, claiming to have witnessed Tobias's mother committing theft and threatening to report her to the police. It is noteworthy that even in this case Todd is twisting the truth: Mrs. Ragg caught him stealing from the house where she worked, and she talked him into giving back what he took and did not report him.[82] Twisting the facts gives them the sound of truth, and Todd manages to seal Tobias's lips. 'You may think what you like, Tobias Ragg, but you shall only say what I like', he states, and Tobias exclaims: 'I will say nothing—I will think nothing'.[83] He is true to his word. When the captain and the Colonel, Thornhill's friends, interrogate the boy, all they get from him is: 'I know nothing, I think nothing', and 'I cannot tell, I know nothing', a frightened 'Nothing! nothing! nothing!' and a final 'I have nothing to say […] I have nothing to say'.[84] Todd has successfully managed to assert control over Tobias: the boy's tongue is bound, his speech is completely annihilated, and he is deprived of volition. The meaning of his own language starts slipping: he claims he 'cannot tell', which means both that he does not know, which is a lie, and that he is not allowed to tell, which is the truth he cannot utter. Deprived of speech, Tobias is utterly powerless. Finally, however, when the awareness that Todd is a murderer becomes too heavy a burden for him, he resolves to tell the police. Then, Todd takes the ultimate step towards silencing him beyond recall: he shuts Tobias in a madhouse, a place where truth, however loudly it is screamed, is never believed.

Confinement, as well as exclusion from knowledge, is the technique Todd and Lovett adopt with Mark Ingestrie as well. Being the cook, he is the person who lives the closest to the truth; therefore, Lovett and Todd's policy prescribes his confinement to the basement, where he is doomed to die, eventually. Although he is a prisoner from the moment he sets foot in the cellar, he is not aware of his situation. Only when he becomes restless does he receive a note that officially informs him of his prisoner status, and the moment Mark is *told* he is a prisoner, he knows it. Again, it is the power of the actual words, though in written form, that makes incarceration true for him. Until then, only Sweeney Todd and Mrs. Lovett knew his true status and deciding when to reveal it was their prerogative. Mrs. Lovett points out as much to Mark when she hires him. All will be well as long as he will be 'industrious', but, should he become 'idle', he will 'get a piece of information which will be useful, and which, if [he is] a prudent man, will enable [him] to know what [he is] about'.[85] What Lovett disguises as an exhortation to industriousness contains a death threat that Mark cannot grasp. The 'piece of information' is that he is no longer master of his own destiny, and that his life is at stake unless he obeys. In a plot that works according to the synonymous nature of knowledge and power, Todd and Lovett sit right on top of the characters' hierarchy, preceding Foucault's theorization of the connection between knowledge and power by more than a hundred years. Their exclusive access to truth, their control over the act of turning it into spoken or written words, assert the murderous couple's power over the other characters. Locking Tobias and Mark away, Todd and Lovett turn them into a lunatic and a prisoner respectively, a change of status which is consistent with Foucault's theory that truth belongs to the outcast of society.

Similar techniques of discourse control and exclusion marked the passing of the Act, and can be summarized in John Abernethy's statement that 'the Act is uninjurious if unknown'.[86] An exemplary instance of how these strategies were deployed is the 'Nattomy Soup' case.[87] In May 1829, a new inmate of St Paul's workhouse, in Shadwell, managed to sneak in a newspaper reporting parliamentary discussion of the Anatomy Bill. He read it to the other inmates, who grew alarmed; then, at mealtime, he voiced his 'suspicion' that the soup might contain 'human as well as animal remains'.[88] The troublesome inmate was sentenced to a House of Correction, which punishment, Richardson underscores, was administered, not on the basis of the accusations he made against the Anatomy Bill, but because he distressed the other inmates and made false claims about 'the workhouse broth'.[89] In short, the court never even mentioned the Anatomy Act. As with Tobias and Mark in *Sweeney Todd*, the troublesome inmate who disturbed the status quo, alerting the other inmates to frightful possibilities and challenging the system by making uncomfortable statements, was isolated, while the topic of anatomy was ignored. In pre-Act England, as in Todd and Lovett's

London, the topics of human dismemberment and cannibalism were taboo.

The taboos Todd and Lovett create highlight the most conspicuous feature of truth in *Sweeney Todd*, which is its being unspoken. A feeling that truth simply cannot be uttered marks passages in the plot in which characters blatantly ignore the truth, even when this is, quite literally, under everybody's nose. This is most evident in the episode titled 'The Strange Odour in Old St Dunstan's Church'. St. Dunstan's congregation is peculiarly 'pious', a word repeated several times in the episode, and which clashes with the hypocritical behaviour of the members. As a 'strange and most abominable odour'[90] slowly but steadily fills the church, people complain and protect their noses with a variety of aromatic contrivances, but otherwise remain peculiarly inactive. While they 'generally agre[e]' that the stench 'must come [...] out from the vaults beneath the church', the problem does not 'acced[e] any reply'.[91] The 'pious and hypocritical Mr Batterwick' reasonably argues that 'the present books' 'satisfactorily prov[e]' that no one has been buried in the vault of late, and therefore he excludes that 'dead people, after leaving off smelling and being disagreeable, should all of a sudden burst out again in that line, and be twice as bad as ever they were at first'.[92] Of course, official records can shed no light on the actual problem: the smell arises from the bodies of Sweeney Todd's victims, which the barber discarded in the vaults under St. Dunstan's. Mr. Batterwick's insistence on the official records mocks the want of firmness on the part of London's authorities—be they governmental or parish authorities—in similar circumstances: as there are no official records of recent burials, nor there is an official reason that should prompt an inspection of the vaults. The narrator explicitly criticizes this dynamic, stating that the problem of St. Dunstan's smell 'began to excite some attention' only after several months, because 'in the great city of London, a nuisance of any description requires to become venerable by age before anyone thinks of removing it; and, after that, it is quite clear that that becomes a good argument against removing it at all'.[93]

As they have found a reasonable objection to intervening, the congregation tacitly makes the smell a taboo topic. Not even the prospect of a visit from the bishop spurs them to action. Indeed, the churchwardens 'flatt[er] themselves, that perhaps the bishop would not notice the dreadful smell, or that, if he did, he would [...] say nothing about it'.[94] The bishop, though, crushes their hopes as soon as he arrives: he immediately overrides the taboo by openly asking about the smell, and he is hardly impressed by the churchwardens' justifications. Hearing their hesitant admission that they are 'afraid' that the 'horrid, charnel-house sort of smell' is always there, the bishop exclaims: 'Afraid! [...] surely you know; you seem to me to have a nose'.[95] By uttering the truth and underscoring the congregation's dissociation from the reality of their sensorial experience, the bishop disrupts the precarious balance established by the silence that hung over the topic of the smell until that moment and prompts the church authorities to action. In the end, they intervene only when their social status is in jeopardy: if the 'frightful stench', the narrator reasons, 'had been graciously pleased to confine itself to some poor locality, nothing would have been heard of it; but when it became actually offensive to a gentleman in a metropolitan pulpit, [...] it became a very serious matter indeed'.[96] Interestingly, before the accident with the bishop, the only action the congregation took was that of 'slinking' into Bell Yard to visit Lovett's pie-shop, and

> relieve themselves with a pork or a veal pie, in order that their mouths and noses should be full of a delightful and agreeable flavour, instead of one most peculi-arly and decidedly the reverse.[97]

The behaviour of St. Dunstan's congregation connects truth as a problematic element in the narrative with the wider context of the Victorian metropolis, its scandals related to burial ground overcrowding, and its social organization. There are fair grounds to suppose that the 'St Dunstan's' episode draws in some measure from the Enon Chapel scandal, which broke in 1844, two years before the first episode of *Sweeney Todd* was issued. Enon Chapel, later known as Clare Market Chapel,[98] was built in Clement's Lane, not far west from the spot in which the action of *Sweeney Todd* is set (Fig. 2).[99]

While the upper floor of Enon Chapel was used for masses, its vault was used as a burial place and, as the fictional St. Dunstan's, emanated a noxious smell. The minister who managed the chapel speculated over burial fees, crowding into the limited space a quantity of coffins far exceeding its capacity. He would remove old (and not very old) bodies to make room for fresh ones, employing cartmen to dismember the remains and move them or flush them away through a sewer that conveniently ran under the vault. Perhaps later he decided he could dispense with the services of the cartmen, as he took to removing the

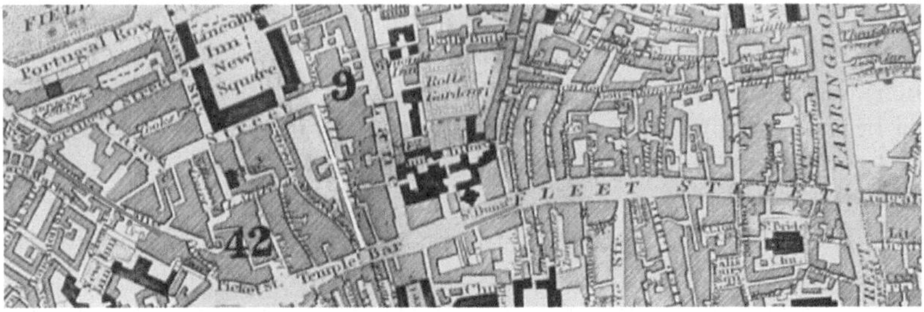

Fig. 2 Greenwood, Christopher and John. Map of London, from actual survey, comprehending the various improvements to 1851, detail. Clement's Lane, on the left is marked in red. Further east, Bell Yard is marked in yellow. Yet further east, in Fleet Street, St. Dunstan's is marked in green. © British Library Board 03/07/2018. Shelfmark: Maps Crace Port. 7.265. Item number: 265

remains himself and burying them under his kitchen, which communicated with the vault through a door. In August 1844, the new owner of the house decided to lower the kitchen floor, as the ceiling of that room was strangely low. The man employed to do the job had a nasty surprise, finding under the upturned flagstones the bones of Enon Chapel's dead. After several Sundays of work, he gave up, finding 'the less destructible portions of this army of dead, although passive in their resistance, "beyond his management"', and the bubble of silence around Enon Chapel finally exploded.[100] Dr. George Walker devoted a considerable part of his campaign against intra-mural burial to denounce Enon Chapel. He widely discussed it in his report *The Grave Yards of London* (henceforth *Grave Yards*), which he presented before the House of Commons, published in 1841, and he addressed it repeatedly in his later writings. In *Second of a Series of Lectures*, which contains a summary of Walker's work on Enon Chapel, he wrote that the 'lower part, kitchen, cellar, or "DUST HOLE"' was 'devoted to the dismemberment and desecration of the dead'.[101] Mr. Burn, the master cartman, bore witness to the offensive state of the chapel and the freshness of some of the bodies he removed. He also declared himself certain that the sewer was regularly used to dispose of the bodies.[102] Whittaker, an undertaker who appeared before a Select Committee, also vouched for the freshness of the bodies removed from Enon Chapel, and he also testified to the use of quick-lime on the bodies to accelerate decomposition.[103] A cabinet maker named Pitts insisted on the dreadful smell of the place, especially over summer, when it became strong enough to provoke headaches.[104] It took sixteen years for the case to be discussed, which confirms *Sweeney Todd*'s narrator's observation that nuisances in London became 'venerable by age' before anyone addressed them, especially if they concerned poor neighbourhoods.

Mismanagement of burial grounds reserved for the poor was widespread: an article from 1846 titled 'Desecration of the Dead' reported the case of the overcrowded burial ground behind St. Giles's workhouse, which was unearthed during the works to enlarge the building. The scene was similar to that Enon Chapel presented: bodies in all stages of decomposition were unearthed, some pits containing as many as '14 [coffins]'. The coffins and their 'ghastly occupants' could 'be traced within 13 or 14 inches from the surface', and the reporter expressed his concern about the 'fearful results to the sanitary condition of so densely crowded a neighbourhood [that] will follow the opening of the loathsome pit now exposed to view'.[105]

The exasperated proximity to death and the dead that characterized the mid-Victorian city did not impact on all classes equally, and poor people's bodies were disposed of differently than the ones of the members of the middle and upper class. Almost every church in the city of London had its own (severely overcrowded) burial ground,[106] which made the sight and smell of the dead commonplace. When new bodies were to be accommodated, the coffins of the poor were disinterred, the corpses broken with a spade and shovelled in a hole dug nearby. Anything that could be recycled, such as the nails, was resold, while the chopped coffins would be used as fire wood.[107] The denizens of poor neighbourhoods were more exposed to the sights and smells of death, and the dwellings surrounding the overcrowded cemeteries were unprotected against pollution from the decomposing matter that saturated the ground. The inferior standards of care that were reserved for the tombs of the poor emerge in the 'St Dunstan's' episode, which shows that the perception of the very real problem of cemetery overcrowding, and its characteristic smell of decay, decreased as the individual's spatial and social distance from poverty increased.

Considering the absence of the figure of the surgeon, which connects truth to sight, and considering the primeval nature of cannibalism, which constitutes the chief discourse in the narrative, it is unsurprising that smell should be the litmus test of truth in *Sweeney Todd*. Smell was a powerful component of Victorian London life, and it influenced its literature. Analysing smell descriptions in novels from the 1860s, Janice Carlisle argues that, not only were smells part of the code through which Victorians constructed class, but also that Victorians, preceding twentieth-century scientific studies on smell, 'accepted as common sense' the fact that smell had 'less to do with thought than with feelings'.[108] They inferred its 'inescapable materiality', de facto sanctioning its condition as a lowlier sense than sight (the sense of the medical man) and hearing.[109] She also observes that the 'stench of the poor', a mixture of 'disease and death', typical of 1840s novels, all but disappears in 1860s novels, which fact she connects to the mid-nineteenth-century process of sanitation and its attempts to 'sanitize and deodorize public and private life'.[110] Indeed, Flanders notes that by mid-century, the destitute had turned into an 'alien race',[111] and were progressively isolated in circumscribed neighbourhoods that were constantly subjected to works of 'improvement' and 'ventilation', that is, they were destroyed to build rich ones.[112] The euphemistic term 'ventilation' is particularly meaningful, considering that the poor neighbourhoods were the places that produced the actual stinks: the powerful slaughterhouse stench,[113] the reek of pigs kept in the house,[114] and the nauseous odour of unmaintained privies,[115] were typical slum smells, as well as the ghastly stench of overcrowd ed cemeteries. Furthermore, as Stephen Halliday notes, according to the miasmatic theory of contagion that prevailed for the best part of the nineteenth century, 'disease was caused by inhaling air that was infected through exposure to corrupting matter'.[116] The conception of poor neighbourhoods as unsanitary spaces extended to the very air that could be breathed within their boundaries.[117] The 'great unwashed' were perceived as a group naturally characterized by stink, which was also considered the chief channel of contagion, and were consequently isolated and subjected to the moral judgement of the middle and upper classes. Therefore, it is unsurprising to hear St. Dunstan's 'pious congregation' expressing concern about the smell of death and decay impregnating their church: the smell of poverty par excellence had moved to their respectable parish, curtailing their spatial and social distance from the poor. Their refusal to acknowledge the presence of something as unrespectable as stink underscores their hypocrisy, and their unresponsiveness to the powerful, but 'lowlier', stimuli that reach their noses allows the murderous Todd-Lovett partnership to operate undisturbed in their neighbourhood. Furthermore, by failing to acknowledge the truth their instinctive olfactory response to the stench suggests them they get involved, though unwittingly, in the crime perpetrated. Their attempts to protect their social status by ignoring, that is, not speaking about, the smell of death and decay in their church aggravate the problem; it is only when the smell is brought into the powerful world of the spoken word that truth finally emerges.

The isolation to which the Victorian poor were subjected, and the moral judgement that accompanied this isolation, facilitated the passing of the Anatomy Act and its sanctioning of the connection between dying in poverty and dissection. The very expression that indicated the eligible bodies in the Act, 'unclaimed', implies distance, separation. Anonymity made it easier for the middle and upper class to relinquish any sense of responsibility for the exploitation of the bodies of the pauper, which on paper was devoted to the higher purpose of the common good, but in reality implemented a service that only benefited the wealthy. The pious St. Dunstan's parishioners complain about the awful smell of decay, though not taking any concrete action about it, and then rush to Lovett's in Bell Yard to feast upon the pies produced with the same matter rotting under their church. Likewise, the middle and upper class alienated the 'great unwashed' and the unappealing smell of their bodies, houses, and neighbourhoods; yet, quite ironically, the same bodies that could not be touched in life were to be touched in death by the apprentice surgeons perfecting their skills in view of exercising them on the patients who could pay for the service.

Although truth is hidden and unspoken, the plot of *Sweeney Todd* tends inevitably towards the final dénouement. Actually, the series tells the story of how the truth surfaces from the underground world of silence and slippages of meaning in which it is kept captive. The series culminates in a powerful upturning of the situation, in which characters speak out loud the taboo discourses. Once the prohibition to speak the truth falls and the words are uttered, truth becomes visible, becomes *true* for everyone. The bishop in the 'St Dunstan's' episode starts this process by overriding the congregation's tacit agreement to make the smell a taboo topic. However, it is Mark Ingestrie who officially breaks the plot's greatest taboo. Throughout the story, the pies are often defined to be 'Lovett's', and they are always 'made *by*' an unknown, unspecified 'someone'. The pies are 'Lovett's', and that is enough for Lovett's customers. As long as this situation is stable, the pies are praised for their taste and delicious smell and the customers salute them as a medicament. This dynamic break when Mark Ingestrie escapes the pie-manufactory through the same windlass that brings the pies upstairs in the shop, keeping the underground world of manufacture out of sight and out of mind. As he reaches the pie-shop, he springs out, like a jack-in-the-box, declaring:

> Ladies and Gentlemen – I fear that what I am going to say will spoil your appetites; but the truth is beautiful at all times, and I have to state that Mrs Lovett's pies are made of *human flesh!*[118]

In this crucial dénouement, the pies are finally made *of* something. Nobody questions the authority of this ragged stranger accusing the best pie-maker in London of unspeakable crimes. He speaks the words and, as in the case of the bishop, the word is all-powerful: once uttered, it alters reality, and the ghastly flesh-pies become real. All of Lovett's customers feel sick simultaneously and spit the 'clinging' portions of the pies.[119] Only afterwards does the policeman who is there to arrest Mrs. Lovett corroborate Mark's statement.

Such a spectacular conclusion fits very well the sensational style of the penny blood genre: a man secretly kept captive, gory deeds, ghastly poisonous food, all combine to create the perfect ending to the *Sweeney Todd* series. The marriage of Mark and Johanna is the comforting happy ending that Crone identifies as typical of penny bloods.[120] Yet, I would see Mark's statement that 'the truth is beautiful at all times' as the narrative's attempt to go beyond the comforting ending, indeed as an attempt to leave readers a prompt for reflection, as an aftertaste, on the importance of knowing the truth. As the troublesome inmate of Shadwell's workhouse believed, knowing the truth, no matter how unsavoury, is crucial to be able to act in one's own best interest, in order not to participate in an involuntary act of cannibalism, neither as the eater, nor as the food.

Throughout the narrative, truth consistently 'emerges', that is, rises from below the ground. The space below the pavement of Bell Yard, St. Dunstan's, and Fleet Street is where the truth is hidden, the place from which it tries to escape. Truth in *Sweeney Todd* inhabits the underground, conforming to the urban configuration of the Victorian city: London's subterranean space was deeply rooted in the mind of its population as the space of darkness, monsters, fear, and truth.

3 The Dreadful Fall: Death and Survival in the Subterranean Space

Contemporary novelist Neil Gaiman observed that, although the details of the different versions of the story of the demon barber change, each version is invariably 'very location specific'.[121] Mack expanded on Gaiman's comment, noting that the narrative displays an obsession with the 'exceptionally heightened and narrow representation' of London's space.[122] All versions and adaptations of the *Sweeney Todd* story consistently limit the action to an area of Fleet Street that encompasses 'Temple Bar, St Dunstan's, Bell Yard, and the Precincts of the Inner Temple to Temple Stairs on the river', and stretches 'along Fleet Street from just beyond Chancery Lane in the west, to Fetter Lane in the east', a degree of geographical precision, Mack observes, that increased the impression of realism in the story.[123] He also notes that Fleet Street was a peculiar spot in the Victorian city: it was the beating heart of the press business and city gossip, was placed in a strategic geographical position where the ships coming up the Thames and the City met, and the 'labyrinthine series of courtyards and alleyways that spread [...] around it' promoted the encounter of people from very different social backgrounds to a degree that was not to be found in other areas of the city.[124] Moreover, this geographical and social intersection had its own market and prison,[125] and was not very far from the infamous Smithfield Market, with the adjacent St Bartholomew's Hospital, as well as from Enon Chapel, as we have seen.

We can better understand how *Sweeney Todd*'s steep, circumscribed geography relates to London's geospace, and consequently better navigate it, by considering David L. Pike's theorization of a 'vertical model' for representing the landscape of the Victorian city.[126] Pike notes the impact of the increasing social and spatial distance between classes on the separation strategies adopted to face the logistic and sanitary challenges posed by the confluence of individuals from different social backgrounds in the space of the metropolis; the result was a new 'vertically divided space' that came to define the shape of the urban space.[127] In this new model, 'the underground' came to be identified with concepts related to non-visible spaces, which in London concretized in a 'discourse' on the 'disposal of waste'.[128] Within this discourse, anything that was perceived as refusal was 'flush[ed] out of sight', including any remains of 'precapitalist social structures', which, Pike contends, instead of being effaced, were empowered by their new outcast condition, both 'as allure and as threat'.[129] While he explains this process partially by embracing Stallybrass and White's psychoanalytic model that reads the subterranean space as society's 'unconscious' and the 'locus of truth', Pike emphasizes that the new vertical structure must not be understood as solely related to the middle class.[130] Any understanding of the nineteenth-century conception of the subterranean space must account, he argues, for two elements: firstly, the 'mythic component of the descent to the underworld in search of truth', and secondly, the lower echelons of society, who experienced filth in a counterintuitive way, forever torn between aspiration to 'middle-class respectability' and 'underground criminality', and who likewise contributed to the 'production of the nineteenth-century city'.[131] In the 1960s, Gaston Bachelard used a similar approach in his vertical model of 'psychological space', represented as a house that polarly opposes the 'rational' space of the attic to the 'irrational' space of the cellar.[132] In Bachelard's model, cellars are the spaces where unsightly sights are chased—the 'buried madness' whose 'marks [...] we prefer not to deepen'.[133] These subterranean spaces are also extremely active: underground, 'secrets are pondered, projects are prepared' and 'action gets under way'. Bachelard views the cellar as peopled with its own darker, mysterious inhabitants and, like Pike, he contemplates a descent of our unconscious in its depth, so that we can explore it 'with a lighted candle' and confront its terrors.[134] By applying Pike's and Bachelard's readings to *Sweeney Todd*'s glaringly subterranean world of cannibalism and dark forces, this emerges as a space in which working-class readers could confront their anxiety about annihilation in a savage underground world, and vicariously concretize (through the characters) their aspiration to emerge from such space empowered by the awareness of their position and the capability of changing it. While in *Varney* and *Manuscripts* the movement between subterranean and surface spaces occurs chiefly upwards, the eminently downward movement of the precipitous fall through the trapdoor in the barber shop characterizes *Sweeney Todd*. This is suddenly overturned by Mark's final, triumphant ascent, through which the reader achieves catharsis. To understand this relationship between subterranean and surface space in *Sweeney Todd*, it is first necessary to consider the geospace of the Victorian city itself and the city dweller's perception of it.

Boundaries between above and below were constantly disrupted in nineteenth-century London. The overcrowd ed cemeteries spilled the content of the graves on the surface, and what did not resurface sometimes leaked through the cemetery walls in the form of liquid decaying matter.[135] Public works contributed to fuel anxieties about this uncanny inversion in the urban space. Wise notes that the 'Metropolitan Improvements' of the 1840s uncovered the eerie hidden roots of the city, increasing the Londoner's 'urban paranoia about subterranean spaces'.[136] This concern pivoted on the popular image of the innocent person, possibly new in town, disappearing in the subterranean world, in which the burgeoning city is pictured as a monster that swallows up newcomers as if absorbing them through the ground. If the dead resurfaced from their tombs, after all, why could not the living take their place under the earth? Sweeney Todd's customers, disappearing through the trapdoor on which the barber's chair is mounted, reiterated this type of anxiety. The trap-door chair fits naturally in an urban landscape that popular lore represented as a place where 'the pathways on the streets [were] full of trap doors which dropd [sic] down as soon as pressd [sic] with the feet and sprung in their places after the unfortunate countryman had fallen into the deep hole [...]'.[137] The poet John Clare, new to London, believed this murky portrait of the city his artist friend Rippingille pictured for him, and behaved accordingly. He kept 'a constant look out', imagining every woman on the street to be a prostitute ready to lure him 'into a fine house were [sic] I should never be seen agen'.[138] This must have been wonderfully entertaining for Rippingille, who, Wise suggests, was probably enjoying teasing his friend.[139] Yet, it is noteworthy that his joke specifically framed the underground space as threatening and voracious.

As Mack notes, Dickens, the narrator of Victorian London par excellence, makes a similar representation.[140] As soon as Tom arrives in London in *Martin Chuzzlewit*, he wishes 'to have those streets pointed to him which were appropriated to the slaughter of countrymen'.[141] Later, when he realizes that he is late for an appointment, he is sure that his friend will think that he has 'strayed into one of those streets where the countrymen are murdered, and that [he has] been made meat pies of'.[142] The narrator, however, reassures the reader that 'Tom's evil genius did not lead him into the dens of any of those preparers of cannibalistic pastry, who are represented in many standard country legends as doing a lively retail business in the Metropolis'.[143] Although it is tempting to read in these lines proof of the demon barber's existence,[144] Mack suggests that they rather attest to the fact that the conflation of the burgeoning and 'rapacious' urban space with fears about annihilation of one's individuality were a substantial component of the 1840s 'urban zeitgeist'.[145] Notably, the fears that according to Mack underpinned the Londoners' relationship with the underground space also included disappearance through ingestion.[146] Rippingille informed Clare that after falling into the trapdoor, the countryman 'woud be robd and murderd [sic] and thrown into boiling cauldrons kept continualy [sic] boiling for that purpose and his bones sold to the docters'.[147] In this very detailed explanation surface the anxieties about subterranean space, cannibalism, and medicine that the city's geospace generated, and which emerge forcefully in the narrative space of *Sweeney Todd*'s London.

The warning contained in Rippingille's murky portrait of London's underground was simple: falling below meant dying and being cooked. *Sweeney Todd* reinforced this concept by representing living characters trapped underground as already dead. The unlucky cook Todd murders in Lovett's basement, the one Mrs. Lovett told Mark Ingestrie was gone to meet 'his very oldest friends', is dressed

> but lightly [...] in fact, he seems to have but little on him except a shirt and a pair of loose canvas trousers. The sleeves of the former are turned up beyond his elbows, and on his head he has a white night cap.[148]

This description resembles that of an underground worker, such as a miner, a sewer worker, or a man working near furnaces (as is the case): the sleeves are rolled up to find relief from the heat, and the cap prevents sweat from pouring into the eyes. However, the emphasis on the looseness and scantiness of the clothes adds a dark undertone to the description, as if this man does not need proper garments because his clothes will be his shroud. The fact that he is wearing a cap, an item that sometimes figured in the stock burial apparel,[149] reinforces this impression. The man in the basement, which is peculiarly 'sepulchral',[150] is already dead to the world, out of reach of anyone who might help him. Not only does the vast, monstrous city trap the unaware underground, but, as Crone notes, it also creates the anonymity necessary to prevent anyone noticing somebody else's disappearance.[151] When Mark Ingestrie starts realizing that he is a prisoner Lovett's sepulchral basement, he wonders:

> [i]s it possible that even in the very heart of London I am a prisoner, and without the means of resisting the most frightful threats that are uttered against me?[152]

Apparently, it was. In 1842, four years before the first episode of *Sweeney Todd* was issued, a Select Committee heard the witness of a workhouse inmate who was unjustly committed to forty-eight-hour imprisonment in a windowless cell, or 'Black-hole', together with five more people.[153] It was August, and the temperature soon grew unbearably hot; when the prisoners complained, the workhouse staff nailed their air hole shut.[154] Such stories fuelled anxieties about anonymity and disappearance in the urban space of the city, where it appeared that the sheer numbers of the population prevented anyone noticing—or caring—if someone disappeared. Surgeon George Guthrie capitalized upon this point in his 1829 open letter to the Home Secretary, in which he protested the allegation that anatomist s were secretive about dissecting-room proceedings. He wrote that the doors of 'every dissecting room in London [were] always open' to the public, but that laypeople did not concern themselves about 'what [was] going on': 'in London', he stated, '[...] no one knows or cares what is going on, unless he is interested in it'.[155] Although Guthrie's claim about the openness, practical and metaphorical, of London's dissection rooms is debatable, as Wise notes,[156] his statements confirm the idea of a burgeoning urban population generally indifferent to what did not concern them personally. Guthrie was right only in part: the existence of cautionary tales indicates that at least a portion of the population may not have been entirely unreceptive, or unconcerned, about what went on in the city's underground space.

Rippingille specifies that one who fell underground would have his bones boiled and 'sold to the docters'. The meat and the pies, and consequently the dead cook in Lovett's basement, stemmed from the idea that had developed in the popular mind during the first decades of the nineteenth century according to which 'the notion of boiling, cooking and consuming had become intermingled with the notion of dissection and anatomy'.[157] In this respect, Wise brings in example the case of Caroline Walsh, an elderly woman who accepted the invitation of a neighbouring couple, Eliza Ross and Edward Cook, to occupy a bedroom in their house. Cook was known to be a resurrectionist, and Walsh's granddaughter, Anne Buton, warned her against accepting the invitation: 'If you go to stay at the Cooks, they'll cook you!'.[158] Soon after she moved in with the Cooks, Walsh disappeared. Buton then started searching for her grandmother and the newspapers took interest in her case, alleging that her grandmother had been 'burked for the base object of selling her body for anatomical purposes'.[159] Twelve days before the 'Italian Boy' case exploded, Eliza Ross was arrested and charged with murder. Her twelve-year-old son stated that his mother had single-handedly smothered the old woman, put her in a sack, and sold her to the London Hospital in Whitechapel. Ross was found guilty of murder and hanged.[160]

It is unlikely the Cooks 'cooked' Walsh's body. Yet, it was never found, and Anne Buton's macabre pun supports Wise's argument that anatomy and cooking were connected in the popular mind. Rumours about dissection rooms favoured this association: intelligence of '[m]ysterious attics, rooms with opaque windows, creatures pickled in bottles, body parts in cooking pots, disappearances, strange goings-on after dark'[161] substantiated the idea that resurrectionists and doctors 'cooked' people. Returning briefly to Rippingille's joke, it is meaningful that the cooking of the unaware countryman happens specifically *underground*. The design of medical education spaces contributed to the elaboration of this concept in the popular mind, as the dissecting room was a distinctively subterranean location. To exemplify the type of glimpse laypeople were given of the spaces of anatomy and dissection, Hurren quotes an 1840 article that appeared in the *Penny Satirist* in which an art student related his experience in certain—unnamed—anatomy schools in the capital during the 1830s.[162] The student explained that '[t]he dissection room was underground and there was a museum of skeletons and hearts, livers, legs and lights upstairs', which made him uneasy, particularly after dark.[163] He then specified that the dissection room 'was not down stairs but down *ladder* [sic]. It was simply a ladder through a species of trap door that we made our descent', and there was

> one room [in which there] were the operators, and in another room a sort of back kitchen with a water pipe and sink, where the bodies were washed. In this sink there was generally a body lying, and the water running upon it.[164]

The bodies, usually 'a dozen', had been 'stolen from churchyards or bought on the sly', and alongside them were 'a number of amputated limbs, such as heads, arms & legs, &c. in various stages of *scientific preparation* [sic]'.[165] As is to be expected, the student was 'always glad when [he] got at the top of the ladder'.[166] This description contains all the elements that appear both in Rippingille's mocking cautionary tale and in *Sweeney Todd*. There is an underground environment, a trap door, and mutilated dead bodies lying in a kitchen, which compose a frightful museum of horrors that seemed to have been conjured out of a nightmare. Moreover, as discussed above, laypeople associated anatomy and dissection with butchery, and butchery, for practical reasons, was partially performed below the ground of the Victorian city, as was dissection.

Discussing the controversial space of Smithfield Market, Wise highlights the show of bloodshed that characterized the spot, as well as the habit of keeping (and butchering) cattle underground.[167] Flanders observes that Clare Market, though small compared to others, hosted twenty-six butchers, who slaughtered several hundred animals each week both above and below the market's ground; next to butchers, there may be a tripe boiler.[168] Boiling was also part of the activities of knackers' yards, where old or diseased horses were killed and butchered to produce cat food.[169] The process of boiling, therefore, was associated with butchery and medicine alike, making the step from anatomy to butchering-boiling-cooking and, henceforth, consuming a short one. Additionally, the secrecy the medical fraternity insisted on keeping about their practices reinforced the idea that something awfully wrong went on in dissection rooms. Butchery and dissection thus fuelled the Londoner's anxieties about the underground, turning it into a space where a good person may disappear to be hacked into pieces and used as an anatomical subject by fiendish doctors and medical students. The dark subterranean space of *Sweeney Todd* hypostasized this self-sustaining set of fears.

The basement of Lovett's pie shop communicates with Todd's basement and with St. Dunstan's vaults. This maze of connected subterranean spaces tapped into the Victorian paranoia that pictured the subterranean space of the city as a labyrinth of tunnels used by criminals to move unseen through the urban space. Wise observes that the London burkers' cottages in Nova Scotia Gardens, later known as 'burkers' hole', were rumoured to be connected to one another through a tunnel.[170] Discussing the dramatic potential of the 'irrational' space of the cellar, Bachelard notes how its walls 'are buried walls [...] that have the entire earth behind them'.[171] The cellar is a buried/burial place and finding oneself in this space means being at one removal from being entombed alive. The increase of fear this situation triggers allows to imagine horrors in the cellar's depths.[172] Bishop and Williams' cottages therefore become 'cellar-spaces', or 'holes', which the collective imagination pictures as a subterranean maze through which the murderers, immediately relegated to the depths of the underworld, can be imagined moving and committing their crimes. This operation simultaneously casts these crimes into the imagined depths of the city, out of sight. Bachelard also notes how representations of underground mazes, especially when the location is a city, reflect fantasies of 'dominating in depth the surrounding cellars'.[173] I would venture that *Sweeney Todd* mingles this concept with the Victorian tendency, noted by Pike, to remove the repulsive-fascinating 'filth' produced by the creation of the new social order to a subterranean space that was both geographical and metaphorical.[174] *Sweeney Todd*'s subterranean labyrinth reflects the equally labyrinthine space above the ground in a way that allowed the reader to project into the imagined space terrors imagined in the geospace of the city. Dissection rooms were brick-and-mortar locations in the urban landscape, although the crossing of their threshold was, not only tacitly forbidden, but also undesirable. Accounts such as the one the art student gave to the *Penny Satirist* pictured the space of the dissection room as the stuff of nightmares, and cases such as that of Catherine Walsh and the 'Italian Boy' fuelled a grim picture that became part of the city's collective memory. The 'nattomy soup' case conveys the terror the perspective of dissection raised and the fear of cannibalism it generated. The story of the demon barber, which is one of annihilation through mutilation and ingestion, includes both the trauma and the recovery from it.

In the basement, the victims of Todd's trap-door-chair are butchered, and their bodies are cut into lumps and steaks. The human flesh is undistinguishable from animal meat, and Mark exclaims: 'I never could tell the pork from the veal myself, for they seem to me both alike'.[175] When asked about the source of the meat, Mrs. Lovett answers: 'that is no business of yours'.[176] Something's, or someone's provenance is irrelevant below the ground: butchery effaces individual identity, exactly as dissection turned the individual's body into an anonymous subject, a piece of meat on the dissecting table. After being cut, the human flesh is cooked in Lovett's pie factory 'beneath the pavement of Bell-yard', where 'gleaming lights seem to be peeping out from furnaces', and a 'strange, hissing, simmering sound' hints at the cooking of pies (or perhaps at unseen horrors whispering in the dark), as a 'rich and savoury vapour' impregnates the air.[177] Although fire, and not water, is the core element of this hellish representation, the markedly underground location, and the 'kitchen' connotation of the environment connects this space to the one the art student described for the *Penny Satirist*. Furthermore, Lovett's customers both relish the pies for their rich flavour, and attribute to them healing powers. One asserts that, since his stomach is upset from overeating, he will have a pie 'to settle' it,[178] while another considers them a good omen for the birth of his child. His pregnant wife 'won't fancy anything but one of Lovett's veal pies [...] to have the child marked as a pie'.[179] This detail was likely meant to shock the reader, who could guess the truth by now and understand that the pregnant woman, and therefore her child, have developed a craving for human flesh. The narrative is suggesting, not only that a new cannibalistic society is developing, but also, and more subtly, that the eater attributes healing properties to the act of cannibalism. Mutilated bodies can cure those who can access the cure they produce. This was a tangible reality in the context of the 1840s, in which the Anatomy Act and its aftermath attested to a recent past of burking, resurrectionism, and (still technically active) body traffic. Specifically, as the 'nattomy soup' incident demonstrates,[180] *Sweeney Todd*'s working-class readers imagined themselves more as potential mutilated bodies, than consumers of the cure.

The circumscription of the action to a limited space that precisely matches the geospace of the city allowed the removal of the fear of the subterranean space to an imagined underground in which horrors can happen. The high degree of geographical precision keeps the reader simultaneously safely distant and dangerously close to the action, to the mechanical chair, to the meat cleaver and the grinder, allowing them to imagine themselves as both the pie and the eater. At the same time, when Mark Ingestrie leaves the claustrophobic, suffocating subterranean space through the windlass, the reader can participate in his freedom, which is both verbal and spatial. Mark's spectacular apparition, theatrically springing up and sending pies flying all around, reminds one of the theatrical device of the *diabolus ex-machina* and contributes to the staged feeling pervading the whole plot, which perhaps favoured its swift adaptation into a theatrical performance. However, Mark is no demon: as Varney, he is a revenant that emerges from the underground world of death and mutilation to haunt the living with the truth. Mark's presence in the world above the ground after his journey in the subterranean world is disruptive, but the narrative also presents it as necessary. Pike argues that the underground can be defined a 'spatial heuristic' where are relegated the unspeakable, 'unpalatable' truths that do not fit in the rational organization and discourses of the world above; the 'vertical framework' thus formed makes the truth visible, but it does not make it real, so that any mysteries that may emerge can be solved 'only in underground commonplaces of plot, never in aboveground apportionment of responsibility'.[181] The conclusion of *Sweeney Todd* disrupts this vertical scaffolding; Mark's upwards movement through the windlass challenges the downwards look, the one-way gaze from the 'mainstream ideology' viewpoint that Pike identifies as the weakness of the vertical structure.[182] Truth leaves the subterranean world, allowing the literally unpalatable truth of cannibalism to meet the rationality of the world above and be solved. To an extent, it is certainly true that, as Crone argues, penny bloods tended to console their readership with the belief in the existence of a 'larger moral order', rather than inciting them to revolution.[183] Nevertheless, I would argue that the catharsis Mark's escape provided may have been meant to alert the readers about the possibility of saving themselves from the cannibalistic dungeon by seeking and learning the truth about their position in the vertical social space.

Todd and Lovett hypostasize a set of fears deeply rooted in the English cultural heritage as well as inherent to the specific historical-geographical context of the Victorian metropolis. As monstrous bodies that draw from the mythical-folkloric figures of the devil and the witch, they synthetize anxieties related to the threat the new order of the industrial era posed to the physical body and the intellectual and spatial freedom of the lower-class individual. The loose physicality of the demon-barber can therefore be read as the physical manifestation of Victorian society's monstrous self: as Frankenstein's Monster is composed of different parts of decomposing bodies, so Sweeney Todd is an assemblage of all the ideas, philosophies, and people that mainstream society discarded in the process enacted by the industrial revolution. Todd's ability to move through the underground and the surface and to put the two worlds in connection, trapping and killing the unaware underground, represent the monstrosity of the system in place: through his huge mouth, Todd assimilates people gone astray, whose absence will not be noticed, and who represent everything that mainstream society (i.e. the middle- and upper-class) find encumbering. Todd and Lovett's narrative bespeaks the dread of annihilation through mutilation and ingestion that certainly stemmed, as Powell, Crone, and Mack note, from the exploitation to which the industrial economy subjected the bodies of the workers. I would argue, however, that it also originated in the threat that dissection, in the sense of breaking the body in its distinct parts, posed to the paupers' bodies both before and after the Anatomy Act.

Sweeney Todd, a post-Act popular narrative, appears in a context in which law itself ratified such physical threat, concretizing concerns related to seizing and controlling power. The narrative embraces this topic through the sensational elements of mutilation and unwitting cannibalism, discussing issues of discourse and power within the safe space of the fictional narrative. Furthermore, anxieties of annihilation and body consumption related to the monstrous new urban space produced within the city's geospace an imagined, equally monstrous underground. The narrative space of *Sweeney Todd* safely removed the experience of mutilation and annihilation to the imagined underground, while simultaneously bringing it closer to the reader through the detailed replication of the city's geospace. I would venture, therefore, that the *Sweeney Todd* narrative, while invading the urban geospace and becoming the glaringly 'London' story Mack and others celebrates, constructed a relatively safe space where the reader could enact and face unspoken terrors, and where taboo topics such as the use of the bodies of the pauper under the Anatomy Act could be tackled.

Although, as Mack observes, not many nineteenth-century narratives can match *Sweeney Todd*'s geographical precision,[184] G.W.M. Reynolds's *The Mysteries of London* certainly shows an equal degree of obsession, not only for the accurate representation of the Victorian city's geospace, but also for the representation of a labyrinthine and threatening underground world of trapdoors and tunnels to match the maze of the grim London slums. In the next chapter, I will explore the world of *Mysteries*, a narrative as rambling as the convoluted streets where middle-class characters lose themselves and meet a monster that almost matches Sweeney Todd's malicious cunning. Perhaps thanks to his trade, which makes him as much a part of the world of the living as of that of the dead, Anthony Tidkins, the Resurrection Man, is as dangerous above the ground as he is below it.

Notes

1. For an analysis of the debate, see Robert L. Mack, *The Wonderful and Surprising History of Sweeney Todd—The Life and Times of an Urban Legend* (London: Continuum, 2007), 145–48.

2. Rosalind Crone, *Violent Victorians—Popular Entertainment in Nineteenth Century London* (Manchester: Manchester University Press, 2012), 170.
3. Sally Powell, 'Black Markets and Cadaverous Pies: The Corpse, Urban Trade and Industrial Consumption in the Penny Blood,' in *Victorian Crime, Madness and Sensation*, ed. Andrew Maunder and Grace Moore (London: Hashgate, 2004), 45–46; Crone, *Violent Victorians*, 189; Mack, *The Wonderful and Surprising History of Sweeney Todd*, 41–43.
4. In ibid., 159–61, and ibid., 170–72, Mack lists, besides the story of Sawney Bean, a few similar narratives from France and even from Italy among the possible antecedents of the London story. The most famous French version, which appeared in the monthly magazine *The Tell-Tale* in 1824, was set in Paris and related the murderous partnership between a barber and pastry-cook. It included also the detail of the dog, which in *Sweeney Todd* belongs to Mr. Thornhill and puts his friends on the barber's trail. The Italian version, instead, was recorded in Anthony Pasquin's *Life of the Late Earl of Barrymore*, 1793. In this version, a murderous pastry-cook in Venice makes pies out of children, dropping their bodies in his cellar through a trapdoor.
5. Edward Sackville Turner, *Boys Will Be Boys—The Story of Sweeney Todd, Deadwood Dick, Sexton Blake, Billy Bunter, Dick Barton et al.*, 3rd ed. (London: Penguin Books, 1976), 42. Turner's emphasis.
6. Jonathan Swift, *Selections from the Journal to Stella, A Tale of a Tub, Personal Letters and Gulliver's Travels; Together with The Drapier's Letters, I; Sleeping in Church; A Modest Proposal* (New York: Doubleday, 1902), 113.
7. Ibid., 120–21.
8. See Elizabeth T. Hurren, *Dying for Victorian Medicine—English Anatomy and Its Trade in the Dead Poor, c. 1834–1929* (London: Palgrave Macmillan, 2012), 28.
9. See Mack, *The Wonderful and Surprising History of Sweeney Todd*, 262–73.
10. James Malcolm Rymer, *Sweeney Todd—The Demon Barber of Fleet Street*, ed. Robert L. Mack (London: Oxford University Press, 2007), 4.
11. Ibid.
12. Anna Gasperini, 'Anatomy of the Demons—The Demoniac Body Dealers of the Penny Bloods,' *Journal of Supernatural Studies* 2, no. 2 (n.d.): 136.
13. Mary Shelley, *Frankenstein*, ed. Marilyn Butler, 2008th ed. (Oxford: Oxford University Press, 1831), 56.
14. Ibid.
15. Rymer, *Sweeney Todd*, 4.
16. Ibid.
17. Ibid.

18. Charles Dickens, *The Old Curiosity Shop*, 1994th ed. (Oxford: Oxford University Press, n.d.), 22.
19. Dickens, *Oliver Twist*, 53.
20. Rymer, *Sweeney Todd*, 12.
21. Gasperini, 'Anatomy of the Demons,' 137.
22. Rymer, *Sweeney Todd*, 30.
23. Ibid., 29.
24. Ibid., 30.
25. For further analysis of the language of cannibalism and desire in Sweeney Todd, see Mack, *The Wonderful and Surprising History of Sweeney Todd*, 25–26.
26. Rymer, *Sweeney Todd*, 30.
27. Ibid., 30–31.
28. Ibid., 31.
29. Mack, *The Wonderful and Surprising History of Sweeney Todd*, 126.
30. Rymer, *Sweeney Todd*, 278.
31. In Lee Jackson, *Dirty Old London—The Victorian Fight Against Filth* (New Haven and London: Yale University Press, 2014), 116, Jackson discusses corpses explosions in nineteenth-century cemeteries, explaining that 'sextons and undertakers' usually '"[tapped]" coffins in church vaults, drilling a hole to prevent them breaking open with explosive force'.
32. Rymer, *Sweeney Todd*, 142.
33. Gasperini, 'Anatomy of the Demons,' 140.
34. Ibid.
35. Rymer, *Sweeney Todd*, 163.
36. Ibid., 264.
37. Ibid., 266.
38. Ibid. Rymer's italics.
39. Ibid., 266.
40. Powell, 'Black Markets and Cadaverous Pies,' 53.
41. The *Times* issues of November 22 and 26, 1831 included two and three articles on the case respectively, while the issue of December 5 included two articles and the advertisement of Egan's edition of the trial.
42. The *Times* issue of December 6, 1831 advertised 'The Poor Italian Boy—A Pathetic Ballad.' Although in Sarah Wise, *The Italian Boy—Murder and Grave-Robbing in 1830s London*, Pimlico 20 (London: Jonathan Cape, 2004), 90, Wise explains that Italian boys were a common sight in London and attracted the pity of the public, considering the time frame and the proliferation of prints and images celebrating Carlo Ferrari, the Italian Boy, after his murder, I would state that, with a fair degree of certainty, the ballad advertised in the *Times* concerned him. See also ibid., 240–46.

43. 'The Murder of the Italian Boy—To-Morrow Will Be Published...,' *The Times*, December 5, 1831, 4.
44. 'An Affair Took Place in the Public Street on Tuesday Night Week, on the Arrival of the Regulator Coach, Which Has...,' *The Times*, 1831, 3.
45. Ruth Richardson, *Death, Dissection and the Destitute* (Chicago and London: The University of Chicago Press, 1987), 102.
46. 'An Affair Took Place in the Public Street on Tuesday Night,' 3.
47. In 'Police,' *The Times*, February 16, 1829, 6, is reported that a man clapped some treacle on the mouth of another who was in the habit of getting him to pay for his drinks and, as his latest request was refused, called him a bodysnatcher. Later, in 'Police,' *The Times*, November 12, 1831, 4, William Burns was fined 50*s*. for scaring a little boy by applying a piece of tarred sack on his mouth. Burns only meant to take a small revenge against the neighbourhood urchins, who liked to torment him. He was harshly reprimanded, as he was 'old enough to know that such practical jokes, at a time like the present, were calculated to rise the most serious alarm'. See also Richardson, *Death, Dissection and the Destitute*, 194.
48. 'On Tuesday Evening, Between 8 and 9 o' Clock, as Charles White, a Young Lad About 13 Years of Age, Was,' *The Times*, November 24, 1831, 2.
49. Wise, *The Italian Boy*, 144.
50. Crone, *Violent Victorians*, 183.
51. 'The Edinburgh Murders,' *The Times*, January 3, 1829, 3.
52. Camden Pelham, *The Chronicles of Crime, or: The New Newgate Calendar: Being a Series of Memoirs and Anecdotes of Notorious Characters Who Have Outraged the Law of Great Britain from the Earliest Period to the Present Time*. Vol. 2 (London: Thomas Tegg, 1841), 182, https://doi.org/10.15713/ins.mmj.3.
53. Anon., *West Port Murders; or an Authentic Account of the Atrocious Murders Committed by Burke and His Associates* (Edinburgh: Thomas Ireland, Junior, 1829), 90.
54. Ibid.
55. Ibid., 354.
56. Pelham, *The Chronicles of Crime, or: The New Newgate Calendar*, 175.
57. See Wise, *The Italian Boy*, 277.
58. 'Police,' 4.
59. 'Murder of Frances Pigburn, and Apprehension of Rhoda Head Alias Williams, as an Accessory,' *The Times*, November 30, 1831, 3.
60. Wise, *The Italian Boy*, 276–77.
61. *The Trial, Sentence and Confessions of Bishop, Williams and May* (London: Steill, B., 1831), 191.

62. Lisa Rosner, *The Anatomy Murders* (Philadelphia: University of Pennsylvania Press, 2010), 24.
63. Powell, 'Black Markets and Cadaverous Pies,' 45.
64. Rosner, *The Anatomy Murders*, 254.
65. 'Edinburgh Murders,' *The Times*, January 1829, 3.
66. 'The Late Horrible Murders in Edinburgh, to Obtain Subjects for Dissection,' *The Times*, December 1828, 4.
67. Mack, *The Wonderful and Surprising History of Sweeney Todd*, 87.
68. Rymer, *Sweeney Todd*, 281.
69. Hurren, *Dying for Victorian Medicine*, 67.
70. Judith Flanders, *The Victorian City—Everyday Life in Dickens' London* (London: Atlantic Books, 2013), 207–8.
71. Anonymous pamphlet 'Smithfield and the Slaughterhouse,' 1847. Quoted in Wise, *The Italian Boy*, 127–28.
72. Rymer, *Sweeney Todd*, 24.
73. Ibid., 96.
74. Ibid.
75. Ibid.
76. Ibid.
77. Ibid., 98.
78. Michel Foucault, 'The Order of Discourse,' in *Untying the Text—A Post-Structuralist Reader*, ed. Robert Young (London and New York: Routledge & Kegan Paul, 1987), 52.
79. Ibid.
80. Rymer, *Sweeney Todd*, 6.
81. Ibid.
82. Ibid., 149.
83. Ibid., 34.
84. Ibid., 89–90.
85. Ibid., 97–98.
86. J. Abernethy, The Dissector October 1827, 27. Quoted in Richardson, *Death, Dissection and the Destitute*, 219.
87. See ibid., 221–22.
88. Ibid., 221.
89. Ibid., 221–22.
90. Rymer, *Sweeney Todd*, 149.
91. Ibid., 151.
92. Ibid.
93. Ibid., 150.
94. Ibid., 153.
95. Ibid.
96. Ibid., 150–51.

97. Ibid., 151.
98. Walter Thornbury and Edward Walford, *Old and New London—A Narrative of Its History, Its People, and Its Places* (London: Cassell, Petter, and Galpin, 1879–85), 31.
99. In Mack, *The Wonderful and Surprising History of Sweeney Todd*, 182–84, Mack lists the episode as one of the possible historical precedents for the narrative of *Sweeney Todd*, and underscores that the episode was part of the '"Shock-Horror" literature of the period,' ibid., 182.
100. George Alfred Walker, *The Second of a Series of Lectures ... on the Actual Conditions of the Metropolitan Graveyards* (London: Longman, Brown, Green and Longmans, 1847), 18. In Thornbury and Walford, *Old and New London*, 31, Walford gives a different version of the ghastly discovery of the 'modern Golgotha', as something that was discovered during the construction of a new sewer under the chapel.
101. Walker, *The Second of a Series of Lectures...*, 15. Walker's emphasis.
102. Ibid., 17.
103. Ibid.
104. Ibid., 16.
105. 'Desecration of the Dead,' *The Times*, October 16, 1846, 7. Especially in pauper's graves, and in the burial pits of workhouses and hospitals, it was common to find multiple coffins in the same pit, which contributed to exposing the corpses of the poor to the danger of being seized by resurrectionists. In the 'Report from the Select Committee for Anatomy' (London, 1828), 72, anonymous bodysnatcher A.B. stated that the bodies 'of poor people buried from the workhouses' were a favoured prize, 'because, instead of working for one subject, you may get three or four'.
106. Flanders, *The Victorian City*, 219.
107. George Alfred Walker, *The First of a Series of Lectures... on the Actual Conditions of the Metropolitan Graveyards* (London: Longman, Brwon, Green and Longmans, 1849), 16. I will examine in greater detail Walker's discussion of the recycling of funerary paraphernalia in Section 3 of the next chapter.
108. Janice Carlisle, 'The Smell of Class: British Novels of the 1860s,' *Victorian Literature and Culture* 2901, no. 1 (2001): 3–4.
109. Ibid., 4.
110. Ibid., 2.
111. Flanders, *The Victorian City*, 182.
112. Ibid., 188.
113. See ibid., 132–33 and ibid., 138–39.
114. See ibid., 208.
115. See ibid., 206–7; Jackson, *Dirty Old London*, 69–70.

116. Stephen Halliday, 'Death and Miasma in Victorian London: An Obstinate Belief,' *BMJ: British Medical Journal* 323, no. 7327 (2001): 1469. For a detailed history of the miasmatic theory, see also Stephen Halliday, *The Great Filth—The War Against Disease in Victorian England* (Chalford: Sutton Publishing, 2007), 52–87.
117. In ibid., 64, Halliday explains that during the 1840s cholera epidemics, the German doctor Max von Pettenkofer developed the 'Telluric' theory of contagion according to which cholera originated from decomposing diseased bodies and faeces saturating the soil and releasing noxious miasmas. This reinforced the connection between stink and disease in the popular consciousness: members of communities visited by cholera were 'all too familiar' with the smell of 'raw sewage and decaying bodies.' Ibid.
118. Rymer, *Sweeney Todd*, 280. Rymer's emphasis.
119. Ibid., 280.
120. Crone, *Violent Victorians*, 182–83.
121. Quoted in Mack, *The Wonderful and Surprising History of Sweeney Todd*, 83.
122. Ibid., 85.
123. Ibid., 89.
124. Ibid., 86.
125. Which were respectively relocated and demolished in 1848. See Flanders, *The Victorian City*, 76.
126. David L. Pike, *Subterranean Cities—The World Beneath Paris and London 1800–1945* (Ithaca and London: Cornell University Press, 2005), 196.
127. Ibid.
128. Ibid.
129. Ibid.
130. Ibid.
131. Ibid., 197.
132. Gaston Bachelard, *The Poetics of Space*, Kindle (New York: Penguin Books, 1964), Chapter 3.
133. Ibid.
134. Ibid.
135. In Flanders, *The Victorian City*, 221, Flanders notes that such was the case in the Portugal Street cemetery.
136. Wise, *The Italian Boy*, 287–88.
137. John Clare, *John Clare's Autobiographical Writings*, ed. Eric Robinson (Oxford and New York: Oxford University Press, 1986), 132.
138. Ibid.
139. Wise, *The Italian Boy*, 137.

140. Mack, *The Wonderful and Surprising History of Sweeney Todd*, 148–49.
141. Charles Dickens, *Martin Chuzzlewit*, ed. P.N. Furbank (Harmondsworth, Middlesex: Penguin, 1968), 568.
142. Ibid., 576.
143. Ibid., 577.
144. Which, as Mack observes, Peter Haining did in his 1993 study. In P. Haining, *Sweeney Todd: The Real Story of the Demon Barber of Fleet Street* (London: Boxtree, 1993), quoted in Mack, *The Wonderful and Surprising History of Sweeney Todd*, 149–50, Haining argued these were obvious hints to Todd's real story and that Dickens was not more explicit out of delicacy towards those relatives of the victims who might have been among his readers. Mack, in ibid., 50, emphatically rejects these speculations, observing that their chief fallacy is that Dickens wrote *Martin Chuzzlewitt* long before the first instalment of *The String of Pearls* was issued.
145. Ibid., 150.
146. Ibid., 40–42.
147. Clare, *John Clare's Autobiographical Writings*, 132.
148. Rymer, *Sweeney Todd*, 93.
149. Richardson, *Death, Dissection and the Destitute*, 20.
150. Rymer, *Sweeney Todd*, 92.
151. Crone, *Violent Victorians*, 186.
152. Rymer, *Sweeney Todd*, 176.
153. Flanders, *The Victorian City*, 172.
154. The episode, discussed by Flanders, ibid., is reported in Walker, *First of a Series of Lectures*, 30. The name 'Black-hole' may bear connection with a 1756 incident: when the ruler of Bengal captured the city of Calcutta, the Europeans who tried to defend it were imprisoned in the claustrophobic Black Hole, Calcutta's prison for petty offenders, and about twenty of them died as a consequence of the imprisonment. See 'Black Hole of Calcutta,' *Encyclopaedia Britannica Online*, 2014, http://academic.eb.com/EBchecked/topic/67952/Black-Hole-of-Calcutta.
155. Quoted in Wise, *The Italian Boy*, 175.
156. Ibid.
157. Ibid., 173.
158. Ibid., 172.
159. Quoted in ibid., 172. 'Mysterious Disappearance', Globe and Traveller 28 October 1831.
160. In ibid., 173., Wise notes a few obscure points in the case, particularly the incongruity of Ross's son claim that his mother sold the corpse to the London Hospital. Walsh disappeared in mid-August 1831, and anatomy courses usually stopped for the summer, since the heat accelerated decomposition. Indeed, in 'Further Examination of the Partes Charged

with the Murder of Mrs. Walsh,' *The Times*, November 4, 1831, 3, Doctors Luke and Hamilton from the London stated that no corpses were brought or purchased in the days indicated by the defendants, nor in the rest of the month. Moreover, the lecturing and dissecting rooms were under repair. Furthermore, in the *Old Bailey Proceedings*, Buton described her grandmother as healthy and strong. Walsh was therefore an old woman with no interesting deformities, a subject hardly tempting enough for a surgeon to undergo the disadvantages of performing dissection during summer.

161. Wise, *The Italian Boy*, 174–75.
162. 'A Dissecting Room,' *The Penny Satirist*, no. 186 (1840): 1.
163. Quoted in Elizabeth T. Hurren, *Dying for Victorian Medicine—English Anatomy and Its Trade in the Dead Poor, c. 1834–1929* (London: Palgrave Macmillan, 2012), 86.
164. Ibid.
165. Ibid.
166. Ibid.
167. Wise, *The Italian Boy*, 133.
168. Flanders, *The Victorian City*, 132.
169. Ibid., 139.
170. Wise, *The Italian Boy*, 280 Wise's analysis of the imagined labyrinthine underground space in Nova Scotia Gardens will be examined in more detail in relation to the *Mysteries of London*, in the next chapter.
171. Bachelard, *The Poetics of Space*, Chapter 3.
172. Ibid.
173. Ibid.
174. Pike, *Subterranean Cities*, 196.
175. Rymer, *Sweeney Todd*, 174.
176. Ibid., 97.
177. Ibid., 92–93.
178. Ibid., 278.
179. Ibid., 278–79.
180. For further information on lower-class protest against the Anatomy Act, see Richardson, *Death, Dissection and the Destitute*, 219–38.
181. Pike, *Subterranean Cities*, 18.
182. Ibid.
183. Crone, *Violent Victorians*, 191.
184. Mack, *The Wonderful and Surprising History of Sweeney Todd*, 89.

List of Works Cited

'A Dissecting Room.' *The Penny Satirist*, no. 186 (1840): 1.

'An Affair Took Place in the Public Street on Tuesday Night Week, on the Arrival of the Regulator Coach, Which Has...' *The Times*, December 29, 1831, 3.

Anon. *West Port Murders; or an Authentic Account of the Atrocious Murders Committed by Burke and His Associates.* Edinburgh: Thomas Ireland, Junior, 1829. https://doi.org/10.15713/ins.mmj.3.

Bachelard, Gaston. *The Poetics of Space.* Kindle. New York: Penguin Books, 1964.

'Black Hole of Calcutta.' *Encyclopaedia Britannica Online*, 2014. Accessed October 20, 2014. http://academic.eb.com/EBchecked/topic/67952/Black-Hole-of-Calcutta.

Carlisle, Janice. 'The Smell of Class: British Novels of the 1860s.' *Victorian Literature and Culture* 2901, no. 1 (2001): 1–19.

Clare, John. *John Clare's Autobiographical Writings.* Edited by Eric Robinson. Oxford and New York: Oxford University Press, 1986.

Crone, Rosalind. *Violent Victorians—Popular Entertainment in Nineteenth Century London.* Manchester: Manchester University Press, 2012.

'Desecration of the Dead.' *The Times*, October 16, 1846, 7.

Dickens, Charles. *Martin Chuzzlewit.* Edited by P.N. Furbank. Harmondsworth, Middlesex: Penguin, 1968.

———. *Oliver Twist.* Oxford: Oxford University Press, 1994a.

———. *The Old Curiosity Shop.* Oxford: Oxford University Press, 1994b.

'Edinburgh Murders.' *The Times.* January 31, 1829, 4.

Flanders, Judith. *The Victorian City—Everyday Life in Dickens' London.* London: Atlantic Books, 2013.

'Further Examination of the Partes Charged with the Murder of Mrs. Walsh.' *The Times*, November 4, 1831, 3.

Gasperini, Anna. 'Anatomy of the Demons—The Demoniac Body Dealers of the Penny Bloods.' *Journal of Supernatural Studies* 2, no. 2 (n.d.): 133–47.

Halliday, Stephen. 'Death and Miasma in Victorian London: An Obstinate Belief,' *BMJ: British Medical Journal* 323, no. 7327 (2001): 1469.

———. *The Great Filth—The War Against Disease in Victorian England.* Chalford: Sutton Publishing, 2007.

Hurren, Elizabeth T. *Dying for Victorian Medicine—English Anatomy and Its Trade in the Dead Poor, c. 1834–1929.* London: Palgrave Macmillan, 2012.

Jackson, Lee. *Dirty Old London—The Victorian Fight Against Filth.* New Haven and London: Yale University Press, 2014.

Mack, Robert L. *The Wonderful and Surprising History of Sweeney Todd—The Life and Times of an Urban Legend.* London: Continuum, 2007.

'Murder of Frances Pigburn, and Apprehension of Rhoda Head Alias Williams, as an Accessory.' *The Times*, November 30, 1831, 3.

'On Tuesday Evening, Between 8 and 9 o' Clock, as Charles White, a Young Lad About 13 Years of Age, Was.' *The Times*, November 24, 1831, 2.

Pelham, Camden. *The Chronicles of Crime, or: The New Newgate Calendar: Being a Series of Memoirs and Anecdotes of Notorious Characters Who Have Outraged the Law of Great Britain from the Earliest Period to the Present Time*. Vol. 2. London: Thomas Tegg, 1841. https://doi.org/10.15713/ins.mmj.3.

Pike, David L. *Subterranean Cities—The World Beneath Paris and London 1800–1945*. Ithaca and London: Cornell University Press, 2005.

'Police.' *The Times*, February 16, 1829, 6.

'Police.' *The Times*, November 12, 1831, 4.

Powell, Sally. 'Black Markets and Cadaverous Pies: The Corpse, Urban Trade and Industrial Consumption in the Penny Blood.' In *Victorian Crime, Madness and Sensation*, edited by Andrew Maunder and Grace Moore, 45–58. London: Hashgate, 2004.

Richardson, Ruth. *Death, Dissection and the Destitute*. Chicago and London: The University of Chicago Press, 1987.

Rosner, Lisa. *The Anatomy Murders*. Philadelphia: Universty of Pennsylvania Press, 2010.

Rymer, James Malcolm. *Sweeney Todd—The Demon Barber of Fleet Street*. Edited by Robert L. Mack. London: Oxford University Press, 2007.

Shelley, Mary. *Frankenstein*. Edited by Marilyn Butler. 2008th ed. Oxford: Oxford University Press, 1831.

Swift, Jonathan. *Selections from the Journal to Stella, A Tale of a Tub, Personal Letters and Gulliver's Travels; Together with The Drapier's Letters, I; Sleeping in Church; A Modest Proposal*. New York: Doubleday, 1902.

'The Edinburgh Murders.' *The Times*, January 3, 1829, 3.

'The Late Horrible Murders in Edinburgh, to Obtain Subjects for Dissection.' *The Times*, December 29, 1828, 4.

'The Murder of the Italian Boy—To-Morrow Will Be Published ...' *The Times*, December 5, 1831, 4.

'The Poor Italian Boy—A Pathetic Ballad'. *The Times*, December 6, 1831, 1.

The Trial, Sentence and Confessions of Bishop, Williams and May. London: Steill, B., 1831.

Thornbury, Walter, and Edward Walford, *Old and New London—A Narrative of Its History, Its People, and Its Places*. London: Cassell, Petter, and Galpin, 1879–1885.

Turner, Edward Sackville. *Boys Will Be Boys—The Story of Sweeney Todd, Deadwood Dick, Sexton Blake, Billy Bunter, Dick Barton et al.* 3rd ed. London: Penguin Books, 1976.

Walker, George Alfred. *The First of a Series of Lectures ... on the Actual Conditions of the Metropolitan Graveyards*. London: Longman, Brown, Green and Longmans, 1849.

Wise, Sarah. *The Italian Boy—Murder and Grave-Robbing in 1830s London*. Pimlico 20. London: Jonathan Cape, 2004.

CHAPTER 5

The Unknown Labyrinth: Radicalism, the Body, and the Anatomy Act in *The Mysteries of London*

Soon after its appearance on the literary marketplace, *The Mysteries of London* by G.W.M. Reynolds (henceforth *Mysteries*) became one of the most popular and most discussed (read, harshly criticized by the cultural élite) penny bloods. Nowadays, it is one of the most widely researched specimens of the genre. Its intrinsically contradictory nature, which reflected on the response it elicited and on subsequent academic criticism, originates in the contradictions of Reynolds himself. Middle-class born, prominent radical and then Chartist, head of a publishing empire, inveterate bankrupt, held in contempt by middle-class commentators and radicals alike, Reynolds's stated purpose in writing the series was that of instructing the working masses on the gap existing between the rich and the poor in London. In a 'Letter to the Industrious Classes' that appeared in the *Reynolds's Miscellany* in 1847, he commented on his own tendency to include current affairs and politics in his fiction, writing that his digressions aimed to enforce 'the necessity of ameliorating the conditions of the industrious masses. I want to see you well educated, and your position also improved'.[1] This idea is consistent with Chartist fiction's stated aesthetic purpose, which was 'to elevate the marginalized and repressed majority of society'.[2] The first *Mysteries* series, on which I focus here, pre-dates the official beginning of Reynolds's career in Chartist politics. Nevertheless, it shows unmistakable signals of the radical ideas Reynolds developed during his stay in Paris between 1830 and 1837, which would finally lead him to Chartism.

A. Gasperini, *Nineteenth Century Popular Fiction, Medicine and Anatomy*, Palgrave Studies in Literature, Science and Medicine,
https://doi.org/10.1007/978-3-030-10916-5_5

It is virtually impossible, and rightly so, to separate *Mysteries* from Reynolds's politics, nor is it possible to overlook the tension between his politics and his successful career in mass-publication. His contemporaries, including a number of radicals, considered him a hypocrite at the best of the times, a shrewd businessman who exploited the animosity of the masses and the principles of Radical politics for economic gain.[3] Still in the 1970s, scholars such as Berridge[4] and Williams[5] read Reynolds's participation in mass-market production as ultimately rooted in bourgeois culture and politics and maintained that his businessman spirit outweighed his radical ideas.[6] A short time later, however, Anne Humpherys proposed an alternative reading of Reynolds's contradictory character, arguing that it provided him with a powerful 'negative capability', a capacity to 'absorb the contradictory impulses and desires of the populace' that allowed him to balance exceptional entrepreneurial success with the ideals of radical politics.[7] More recently, Haywood emphatically rejected as simplistic the view of Reynolds as a bourgeois posing as a revolutionary for economic gain. He underscored the force of Reynolds's radical message and the presence of bourgeois villains in his stories, as well as the fact that melodrama was 'by no means a class-exclusive discourse', but one that appealed to a readership encompassing individuals of diverse social extractions, which also included 'genteel' women.[8] My purpose here is not that of examining the traces of Reynolds's Chartism in *Mysteries*, but rather focusing on how his political aesthetic interlaced with, and discussed, issues related to the Anatomy Act and the disposal of paupers' bodies. The narrative does not mention the Act explicitly, nor was it explicitly mentioned in Chartist protests, because the target of indignation was rather the New Poor Law. Yet, Richardson explains, this actually included the Anatomy Act as it was perceived as an 'advance clause' of the New Poor Law, and the two were inseparable in the popular mind.[9] In this chapter, I examine the passages in the series that concentrate on issues related to the Act as a piece of clockwork of the great injustice machinery Chartist fiction attacked. After all, one of the most prominent characters of *Mysteries* is a resurrectionist.

The plot of volume I of *Mysteries* is long and convoluted but revolves around a handful of main characters. The hero, Richard Markham, is a young man from an impoverished middle-class family. His rebellious brother Eugene leaves the family house after quarrelling with their father, although he promises to come back to Richard in twelve years, by which time, he assures him, he will be a rich man. A few years later,

Richard, now orphaned and alone, leaves his secluded life in the suburbs for the city, where his naïveté makes him easy prey for a few fashionable gentlemen with a penchant for gambling. In their circle, Richard meets elusive and charming Walter Sidney, a youth who seems desirous of becoming his friend. Walter is really Eliza Sidney, posing as her deceased brother to comply with her benefactor's plans to gain her access to her inheritance. Of course, the 'benefactor' is scheming to seize Eliza's money, and the plan is more illegal than the girl has been led to believe. Her seclusion and disguise expose her to danger: while roaming alone in the streets in the first period of her life in disguise, she gets lost in the terrible neighbourhood of Smithfield, where she is almost killed by criminals. Later, she becomes the target of the lustful plans of a certain George Montague, who participates in the self-proclaimed benefactor's scheme. Montague's name recurs in the story wherever a fallen woman or a dilapidated family appears, and the narrative clearly hints that he is no other than Eugene, Richard Markham's rebellious brother, who is building his wealth by ruining anyone who crosses paths with him.

Because of their alleged friends' machinations, both Richard and Eliza serve time in Newgate prison. Here, Richard meets Anthony Tidkins, alias, the Resurrection Man. This cadaverous, ominous character foreshadows difficult times ahead for Markham. Not only is he a professional bodysnatcher, but he is also a housebreaker and a burker. When they meet again after they have both left Newgate, the young man is struggling to rebuild his reputation and is harbouring hopes towards the beautiful Italian Countess Isabella Alteroni. Markham makes the mistake of antagonizing Tidkins, for fear he should taint his name. This unhappy move makes the Resurrection Man Richard's sworn enemy, a threat to both his reputation and his life throughout the series.

On one occasion, Tidkins' attempt to kill Richard is thwarted by Ellen Monroe. Ellen is the daughter of Richard's guardian, who coincidentally was indirectly responsible for Richard's ruin, as well as his own, through bad investments suggested to him by Mr. Montague. Ellen and her father end up in the ironically named Golden Lane, a miserable slum in which the girl strives to earn a living as a seamstress and becomes prey of a procuress, who aims at making a courtesan out of her. She finds for Ellen odd jobs that progressively impair the girl's modesty while precipitating her from moments of economic independence to periods of dreadful poverty, until the girl accepts the card of a gentleman (Mr. Montague). Richard rescues Ellen and her father from poverty, and Ellen becomes a dancer, as well as a gender-shifter in the same way Eliza Sidney is. In her male disguise, she follows Richard and saves his life.

Adopting the narrative's political agenda as the background of the analysis, I here explore how *Mysteries*' characters and spaces are deployed to discuss corpse disposal and dismemberment in mid-Victorian London.

1 Mephistopheles, Faustus, and Ghouls: Monsters of Anachronistic Injustice

Mysteries' urban labyrinth is crowded with dangerous characters. While some of them, such as George Montague, are mischievous, others are distinctly monstrous in nature. The Resurrection Man and his gang, and Mr. Jones, the gravedigger of Globe Lane cemetery, are creatures whose deformed soul surfaces in their eerie physicality and beastly habits. Their monstrosity also impacts to an extent on the surgeons who purchase bodies from Tidkins. This triad of characters, bodysnatch ers-gravedigger-surgeons, constitutes the propelling force of one of *Mysteries*' chief sensational components, that is, the displacement, dismemberment, and commodification of corpses. Notably, they replicate in the narrative the group of figures that managed the body traffic in the pre-Anatomy Act era: surgeons created the demand, bodysnatchers provided the commodity, and gravediggers accepted bribes from bodysnatchers to grant them access to burial grounds and steal corpses undisturbed. The core feature of this triad of characters, therefore, is its anachronism.

Volume I of the first series of *Mysteries* run between 1844 and 1845, about twelve years after the passing of the Act, which was celebrated as the measure that ended the body traffic. As we have seen, however, matters were different in practice. The convoluted wording, as well as the assumption that it would impact on the pauper's body only,[10] made the Act all but an official authorization of the body trade, which became apparent with the *Rex versus Feist* scandal at the end of the following decade.[11] In 1858, Albert Feist, master of Newington workhouse, was tried for selling a considerable number of 'claimed' dead inmates—that is, people whose 'nearest known relatives', as the Act stated, had stepped forward—to different London anatomy schools with the complicity of Mr. Hogg, the local undertaker. Feist deliberately exploited the Act's ambiguous language and did not clarify for the inmates and their families that 'accepting a pauper funeral meant agreeing to dissection and dismemberment'.[12] He orchestrated fake funerals for the mourning relatives, while he sent the inmate's body to the hospitals; of course, he made up excuses to prevent families from ascertaining the identity of the corpse in the coffin before the lid was nailed down. He was eventually acquitted when it became obvious that carrying the investigation any further would compromise the anatomists. Hurren argues that *Rex versus Feist* proved that neither the Anatomy Act, nor the New Poor Law, secured 'the principle of consent' as 'a bereaved pauper's civil right'.[13] Considering this discrepancy between the theory and the practice of the Anatomy Act, I would suggest that *Mysteries* deliberately deployed the anachronistic monstrous triad to underline the effective continuity between pre- and post-Anatomy Act era in the management of the disposal of the pauper's bodies.

The Resurrection Man of *Mysteries* is one of the most outstanding villains ever produced in the penny blood genre. His name, almost as much as Sweeney Todd's, has become synonymous with the genre, although his character is more complex. Unlike Todd, Anthony Tidkins has a past as a humble youth with aspirations to respectability; society, however, rejected him and brutalized him to the point that he became the fiendish Resurrection Man. Tidkins' ominous sobriquet characterizes him, connecting him indissolubly to death, poverty, and medicine. The definite article '*the*' before his nickname makes him the epitome of the resurrectionist. Everything about him speaks of bodysnatching. The reader sees him for the first time through the eyes of Richard Markham, in Newgate prison. The description takes a mere few lines, and yet, they are enough to create an eerie feeling around the 'mysterious' stranger: he is a 'very short, thin, cadaverous-looking man, with coal-black hair and whiskers, and dark, piercing eyes half concealed beneath shaggy brows of the deepest jet'.[14] As I have noted elsewhere,[15] the Resurrection Man's 'cadaverous' look is his trademark, it is part of his essence.[16] Furthermore, Richard immediately notes that he tends to avert his gaze when he speaks, 'as if he could not support the glance of the person whom he addressed'.[17] In *Mysteries* as in *Sweeney Todd*, the eyes define a character's personality. I have pointed out how Todd and Lovett's eyes reveal their monstrosity, and Anthony Tidkins' eyes are as elusive as Todd's 'squinting' gaze. Whereas the Resurrection Man's eyes are neither misshapen, nor overtly diabolical, they are disquieting. They are piercing, inquisitive eyes that see more in the interlocutor than he would like to show, while at the same time they do not return the look. Anthony Tidkins' averted gaze is a message: there are things of himself that this man does not wish to share.[18]

By contrast, the Resurrection Man is upsettingly frank. After Richard accepts to doing him a favour, Anthony Tidkins introduces himself: '[I]f I can ever do you a service [...] you may reckon upon the Resurrection Man'.[19] The 'ominous title' startles Richard, but Tidkins explains, matter-of-factly:

> 'Yes – that's my name and profession [...] you may know me as Anthony Tidkins, the Resurrection Man.'
>
> 'And are you really –' began Richard with a partial shudder [...]
>
> 'A body-snatcher? [...] of course I am.'[20]

As I noted elsewhere, Tidkins introduces himself by his nickname first, and then by his Christian name because resurrectionism is 'his name and profession', which means it is his nature.[21] In this respect, he challenges the figures of bodysnatchers that emerge from historical records. From the biography of the bodysnatcher James May in the appendix to the *Trial, Sentence, and Confessions* of the London burkers, it appears that May suffered the consequences of being too open about his line of business. Early in his career, he boasted about the money he made from corpses raised from the burial ground of Portugal Street, behind St Clement Danes' workhouse. Soon, he realized that he was 'detested and despised by every person'.[22] His community cast him out, as if his profession was a catching disease, and he moved to another part of the city, where he met Bishop and Williams, according to the biography. Manifestations of hatred against bodysnatchers could be violent. An example is the accident of Holywell Mount Cemetery in Shoreditch, recorded by Bransby Cooper in the *Life of Sir Astley Cooper*: Mr. Whackett, sexton-gravedigger in the private burial ground of Holywell mount, was attacked by a mob after Hollis and Vaughan, two bodysnatchers to whom he had refused access to the cemetery, denounced his connivance with two rival resurrectionists in a police court full of people. When an enraged mob reached the cemetery and, digging up the graves, found some of them to be empty, they almost buried Whackett alive in one of the pits.[23] Professional bodysnatchers, therefore, would obviously tend to secrecy, and they would never assume a nickname as revealing as 'the Resurrection Man'. By contrast, Anthony Tidkins, the literary epitome of the bodysnatcher, makes sure to impress those he meets by exploiting the horror that his name (and profession) raise. His straightforwardness is double-edged: Richard is made aware of what kind of individual Tidkins is, and that, for now, they are in good terms. However, his horror at Tidkins' frankness warns the reader: the bodysnatcher's amiability will not last.

Tidkins' physical monstrosity concretizes Victorian anxieties about the physical and moral decay that proximity to death could provoke: he is a resurrectionist, a house robber, and, of course, a murderer. As we have seen, medical historians note the concern of the nineteenth-century public about dissection's effects on students.[24] Documented evidence of the light-hearted, sometimes downright sacrilegious behaviour of gravediggers and other burial ground staff, such as Walker's reports and lectures, substantiated these fears.[25] These reports also tapped into the annexed concern that constant exposure to the death and decay of urban cemeteries, with their dangerous miasmas, not only did impair the neighbouring denizens' health but also their morality. Anthony Tidkins embodies the idea that the experience of physical decay might cause the decay of the soul: the bodysnatcher's constant physical contact with death dehumanized him both inside and outside. Early in the narrative, he proves himself a selfish, vindictive, and thoroughly detached character, dominated by anger and revenge, a portrait that reiterated the idea of bodysnatchers as thoroughly corrupted individuals diffused among all strata of Victorian society.

Tidkins' construction as the epitome of a bodysnatcher encompasses all the attributes that defined this figure in the Victorian popular mind. The resulting portrait is, conforming to the style of penny blood fiction, exaggerated. For instance, in the early stage of their acquaintance, Tidkins informs Markham that, when he cannot find any work as a bodysnatcher, he does 'a little in another line'.[26] When Richard asks what that 'other line' is, the Resurrection Man graces him with one of his rare full-faced looks and answers evasively: 'Crankey Jem on t'other side [of the wall] will tell you if you ask him'.[27] The other line of business Tidkins is referring to is housebreaking, and it is not surprising that he should be less inclined to talk about it than about resurrectionism: bodysnatching was considered a misdemeanour, because stealing a corpse was not theft, technically speaking. Housebreaking, though, was a different matter. As A.B. stated in front of the Select Committee for Anatomy, thieves caught wandering at night would often pose as bodysnatchers in order not to be prosecuted,[28] which resulted in the assumption that all bodysnatchers were also thieves. A.B. expressed professional contempt for this idea, as he perceived himself to be a very distinct individual from the 'petty common thieves' that were 'ruining' the 'business'.[29] Yet, in *Mysteries*, Anthony Tidkins the Resurrection Man both raises subjects and resorts to thieving when demand for bodies is scarce, tapping into the popular idea that the same category of people committed both crimes.[30] The inclusion of thieving in the Resurrection Man's curriculum, therefore, appears as an attempt to portray a 'real' bodysnatcher, indeed, the quintessential bodysnatcher. Of course, such a representation must include murder.

In 'The Mummy',[31] one of the most sensational episodes of the series, Richard Markham ends up in the Resurrection Man's lair in Bethnal Green, and discovers that Tidkins is a burker, who hangs his victims upside-down in a tub full of water until they drown and then sells their bodies for dissection. As mentioned in the introduction, Wise argues that the character of Anthony Tidkins bears obvious connections to Bishop and Williams, the London burkers: he lives in the same area of Bethnal Green where they lived and his modus operandi closely resembles that of the London burkers.[32] The portrait of the arch-bodysnatcher is thus complete: Anthony Tidkins, the cadaverous-looking man, is simultaneously resurrectionist, thief, and murderer. He certainly is a superlative individual: none of the nineteenth-century professional bodysnatchers of whom we have record was *also* a burker, *and* a thief.[33] Still, the narrative adds yet another dark layer to the Resurrection Man's portrait by suggesting that Anthony Tidkins might not be entirely from this world, that he might be indeed other-worldly, a supernatural creature.

Disinterring dead bodies and bringing them back to the world of the living, where they do not belong, the 'Resurrection' Man, as with Sweeney Todd and Varney, inverts the natural movement from life to death. His practices dehumanize him, turning him into something else, which shows in his physical aspect, as well as in his behaviour. In the famous episode 'The Body-snatchers',[34] where the Resurrection Man and his gang perform the first bodysnatching expedition in the series, the gang's silent communication frames them within a primeval, animal dimension.[35] Tidkins forces open the door of the Shoreditch church where the body they are looking for is buried, while his accomplices and the surgeon who commissioned the body theft lie in wait nearby. At one point, 'the Buffer throw[s] himself flat upon his stomach, with his ear towards the ground' and finally 'utter[s] a species of low growl as if he were answering some signal that caught his ears alone', which indeed is the case: 'The skeleton-key won't open the side-door, the Resurrection Man says', he informs the others.[36] Passages such as this represent the bodysnatchers as beasts: the Buffer's procumbent position, in direct physical contact with the cemetery's earthy ground, signifies his debasement to a lower station, closer to the animal state.[37] This feeling of an animalistic component in his person increases when he reacts the Resurrection Man's inaudible message with 'a low growl'. The same component resurfaces in 'The Exhumation' episode,[38] in which Tidkins and the Buffer disinter a body in Globe Lane cemetery. Their eyes, the narrator says, 'had become so habituated to the obscurity of night, in consequence of the frequency with which they pursued their avocations during the darkness [...] that they were possessed of the visual acuteness generally ascribed to cats'.[39] The eyes, the core characterizing element, in the bodysnatchers are peculiarly inhuman; they are feline eyes that grant night vision, which suggests that their sacrilegious behaviour is maiming their souls, turning them into supernatural creatures.[40] The arch-bodysnatcher Anthony Tidkins, of course, possesses even more strikingly over-developed senses. In 'The Body-snatchers', once the tomb is open, the surgeon explains that they are looking for an elm coffin covered with black cloth. Tidkins then penetrates with an iron rod the lid of a coffin into the vault, tastes the tip and declares, 'smacking his lips': 'Yes [...] the coffin in this vault is an elm one, and is covered with a black cloth'.[41] As if he suffers from a bizarre colour-taste synaesthesia, the Resurrection Man can perceive colours with his tongue. His activities familiarized him with death to the point of modifying his senses, making him the preternatural product of prolonged sacrilegious contact with material decay.

The sense of something sacrilegious about the resurrectionists is conveyed to the reader through the eyes of the surgeon in 'The Body-snatchers'. As he watches the gangs' eerie silent communication as they work to open the tomb, the surgeon

> [cannot] altogether subdue certain feelings of horror [...]: the methodical precision with which they perfor[m] their avocations – and the coolness they exhibit[t] in undertaking a sacrilegious task ma[kes] a powerful impression upon his mind [...] his feelings of aversion were the same he would have experienced had a loathsome reptile crawled over his naked flesh.[42]

The sensory stimuli the image of the 'crawling loathsome reptile' conveys recall to the reader's mind the image of the Buffer prone on the ground, synthesizing it with that of 'the biblical Serpent'.[43] The text suggests that proximity to death affected the bodysnatchers, effacing their reverence for the dead, and relegates them to the category of hardened sinner, beyond pardon or redemption. The idea of supernatural, demonic dehumanization resonates also in Pelham's description of Burke in *Chronicles of Crime*, published only four years before the first episode of *Mysteries* was issued. Pelham dehumanizes and bestializes Burke in the opening paragraph, describing him as '[d]evoid of all sense of humanity' and 'a butcher of the human race'.[44] Furthermore, towards the end of the narrative, as a jail officer interrogates Burke while he is awaiting death, his story takes a moralizing turn: he allegedly affirms that, while in Ireland, he was 'under the influence of religious impressions' and that he used to read 'his Catechism and his Prayer-book'.[45] When asked to explain how he could make such a turn for the worse, he replies

> that becoming addicted to drink, living in open adultery, and associating continually with the most abandoned characters, he gradually became hardened and desperate, gave up attending chapel or any place of religious worship, shunned the face of a priest, and being continually familiar with every species of wickedness he at length grew indifferent to what he did, and was ready to commit any crime.[46]

While the historical accuracy of this conversation is debatable, the moral tale ascribed to Burke's own voice produces a portrait of the burker that is close enough to Reynolds' representation of the bodysnatchers/burkers in *Mysteries*: it pivots on the same concept, that is, evil is catching like a disease and constant contact with decay, moral and physical, will provoke the decay of the individual. The tale emphasizes Burke's estrangement from religion, an estrangement that is spatial as well as moral: he shuns both the places and the people who represent religion, which gives his character a distinctively 'unholy' aura that resonates in the demonic characterization of Antony Tidkins and his gang.

Monstrosity defines the bodysnatchers of *Mysteries*, especially their supreme representative Anthony Tidkins. Considering that the Resurrection Man shows a certain resilience in escaping death,[47] I would venture that Tidkins, the quintessential bodysnatcher, embodies the monstrous contradiction implicit in his profession, that is, the unnatural subversion of movement from death to life. He does not belong to the world of the living, nor to the world of the dead. Anthony Tidkins is a monster in limbo. This liminality extends to his relationship to the time of the narrative and the time of the intended reader. As noted above, his anachronistic presence in a post-Anatomy Act narrative draws attention to the grey area that was the disposal of the pauper's body after 1832. Additionally, and significantly, Wise observes that the invention of a literary bodysnatcher suggested that 'resurrection [...] remained a potent folk memory'.[48]

Considering the time in the narrative in relation to the time of the intended reader, it becomes apparent that the terror and threat the monstrous resurrectionist embodies defy time constraints. The story begins in July 1831; then, there is a flash-forward to 1835. Finally, events progress until they almost reach the intended reader's time, that is, the 1840s. Hence, not only does the narrative annihilate the distance in space from the reader through the setting, which is not simply 'London', but specific areas related to dissection and the anatomy murders, but it also progressively curtails the time distance between the reader and the monster through the series of flash-forwards. Consequently, the Resurrection Man becomes a monster for the almost-present, close to the reader in both time and space.

Sara Hackenberg argues that the Resurrection Man, as Varney, symbolizes the haunting past: Tidkins is 'the inescapable return of personal and political history',[49] the haunting double of the middle-class protagonist, infused with 'republican energy'.[50] I would expand Hackenberg's political reading by considering the Resurrection Man's tendency to survive, which makes him a revenant, as are Varney and Mark Ingestrie. As illustrated in the previous chapters, revenants in these narratives reveal underlying, unpalatable truths. The Resurrection Man's anachronistic figure powerfully symbolizes the injustice of the post-Anatomy Act situation concerning medicine and displaced corpses. He brings the past era of burking and resurrectionism dangerously close to the present era of (alleged) legal order, exposing the Anatomy Act as an ineffective solution. The Act failed to bring the black market for cadavers to an end and, acting in lieu of the bodysnatcher and the burker, sanctioned the anatomist's scalpel's authority over the pauper's body. In this way, thirteen years after the passing of the Anatomy Act, the bodysnatcher becomes a literary monster, the symbol of the haunting past in which the ignominy of dissection threatened the working-class individual. His monstrous presence bears an indissoluble connection with the medical world, and therefore, when *Mysteries* introduces the first medical figure in the narrative, the surgeon of 'The Body-snatchers', he already occupies an ambiguous position that Anthony Tidkins' presence charges with negative connotations.

The body Tidkins and his gang resurrect in 'The Body-snatchers' is that of a middle-class girl. The tomb, an elm coffin, covered with a black cloth, and buried in a vault closed with a marble slab inside a church, is revealing of the social standing of the girl. This was not the bodysnatchers' typical tomb of choice: A.B. explained that they preferred the shallow pauper's graves.[51] In part, the marble tomb can be explained with its intrinsic sensationalistic potential. Its opening proves a long and delicate task that allows the narrator to indulge in a minutely detailed description of the procedure, building up the tension to the climax, when, as was customary, the ruffians manhandle the girl's body, strip it naked and thrust it in a rough sack, tied head to heels. This image introduced a voyeuristic component that would satisfy the audience's demand for cheap sensation, and simultaneously tapped into concerns related to the medical gaze on the female body: Powell observes that the image implicitly suggests that 'the snatching and dissection of the female corpse constitutes rape'.[52] However, I would suggest that the choice of a middle-class tomb, and hence a middle-class corpse, exploits the voyeuristic component of the female corpse to problematize, not just the gaze, but the ethical position of the surgeon in the scene: differently from the counterintuitive way the same issue is addressed in *Manuscripts*, *Mysteries* openly questions the abuse the surgeon commits in opening the tomb.

Montgomery's early study on the relationship between surgeons and bodysnatchers in the Victorian era defined the bond as 'a Faustian bargain'.[53] I argued elsewhere that this definition automatically placed the resurrectionists in the opposite role of Mephistopheles, and that the Victorian medical fraternity capitalized on this juxtaposition.[54] Sir Astley Cooper's statements before the Select Committee for Anatomy exemplify this attitude. He claimed that having to resort to the services of the bodysnatchers was hurtful to the surgeons as 'men of character and education',[55] a statement that framed the resurrectionists as the antagonists of an otherwise honest system, and the surgeons as their victims. To say that this representation was inaccurate would be an understatement. *Mysteries* seems, at first, to espouse this idea, but, as Hackenberg argues, the Resurrection Man, like Varney, 'disrupt[s] any easy categorization of virtue and vice',[56] which also affects the surgeon's position as a 'virtuous' character: a deal requires two parties, and there is no Mephistopheles without a Faustus.[57]

The surgeon in 'The Body-snatchers' is a Faustian figure. Although he is a family friend to the dead girl, he 'called into action' demonic forces to quench his 'thirst after science',[58] because, as with Chillingworth in *Varney*, his 'earnest desire' to open the body overtakes his human feelings.[59] Detachment, apparently, was a taint from which the medical fraternity had not yet redeemed itself thirteen years after the passing of the Anatomy Act: the science-obsessed surgeon was as much alive in the popular mind as the resurrectionist, to the extent that *Mysteries* represented him at the scene of the sacrilegious act. Historical records suggest that this was a rare occurrence, which only characterized the early stages of professional resurrectionism.[60] The surgeon's perception of the resurrectionists strikingly contrasts with his involvement in the snatching. In his eyes, the bodysnatchers become demonic entities speaking their own alien language, as he places himself in the role of a passive observer, similar to the surgeons who testified in front of the Select Committee. Yet, the expression 'called into action' questions his contemplative position. As Faustus, he had the active role of summoning agent, someone who deliberately sought the help of famously dark forces. His Cooperian distress before the bodysnatchers is not credible. While the forces he unleashed terrify him, he is unable, or rather, unwilling to dispense with their services.

His professional figure appears increasingly impaired as the episode progresses. He reveals that the girl's family explicitly forbade him access to her body, which makes her, technically, 'claimed'. Her 'nearest known relatives' paid for her funeral and made an official statement forbidding the dissection of her remains. The doctor overrides the family's (that is, his friends') legal rights and hires the bodysnatchers to steal the corpse. He urges the Resurrection Man to cancel all traces of the violation because, should it be discovered, '[s]uspicion would immediately fall upon [him]; for it would be remembered how earnestly [he] desired to open the body, and how resolutely [his] request was refused'.[61] This detail characterizes the surgeon, not simply as detached, but as a criminal breaking the law. As Chillingworth in *Varney* and Musgrove in *Manuscripts*, he prioritizes his own scientific endeavours over everything else. His connection to the girl's family increases the impression of something callous about him, representing the medical figure as simultaneously untrustworthy and unethical. When he describes what coffin they should haul up from the tomb, he remarks that he has this information because he 'gave the instructions for the funeral [himself], being the oldest friend of the family'.[62] The Faustian doctor, therefore, adds to the component of the haunting past by projecting the idea of a medical fraternity that is still the detached, career- and corpse-obsessed group whose 'thirst' for dead bodies triggered the black market for cadavers and the anatomy murders.

The events unfolding in the following decade confirmed, to an extent, this idea. In 1858, in a letter to the *Times* entitled 'The Difficulties of Dissection', the anonymous correspondent R.H.M. commented on the *Rex versus Feist* case. While he deplored Feist's actions, he observed that the case highlighted the problem of body shortage for anatomy schools and stressed that it was 'a very serious matter to the public' that the supply should be so scarce as to compel the schools to resort to bribery.[63] Therefore, as late as 1858, the medical fraternity still complained that the current legislation, consisting of the combined force of the Anatomy Act and the New Poor Law, did not provide enough subjects for the anatomy schools, fulfilling the early prophecy of 'The Body-snatchers'. Later that year, in October 1858, the provisions of the Medical Act, which envisaged stricter rules for achieving medical qualification in Great Britain and Ireland and limited the power to confer licenses to a handful of (powerful) institutions in the territory, further boosted the demand for bodies without improving the regulations for sourcing them.[64] Already in 1844–1845, *Mysteries*, a popular narrative for the working class, implied that the current legislation did not satisfy the medical fraternity's requests, which, if unheard, might lead to resuming the body traffic, leaving the population exposed. As *Rex versus Feist* would prove, this was the case. The narrative, through the Faustian surgeon, called for an admission of responsibility on the part of the medical fraternity in the anachronistic post-Anatomy Act situation. Even after the Act was passed, the threat of a barbaric past era of burking and resurrectionism hovered over the citizens of London because the surgeons still needed (wanted?) more than they were allotted.

The second bodysnatching expedition the Resurrection Man undertakes introduces the third component of the triad of individuals traditionally associated with resurrectionism, that is, a member of the burial ground staff. The resurrectionists' target is the body of an unknown man who died in the Globe Lane area, and the narrative follows the theft from its commissioning by a private anatomist to the delivery of the body. Mr. Jones, the gravedigger of Globe Lane Cemetery, is crucial to the actual snatching phase of the plan and appears in the episode appropriately titled 'The Gravedigger'.[65] Similarly to the bodysnatchers, he displays several monstrous traits: like them, his work contaminates his body and soul, making him an agent of the system that protracts the gap between social classes beyond death, and he concretizes Victorian anxieties about proximity to death and decay. His most outstanding monstrous feature, which singles him out as particularly hideous, is a consequence of this proximity: unlike any other monster in the narrative, he is a necrophage.

The Globe Lane burial ground, where Mr. Jones works, is a sickening spot. The 'damp' soil emanates a 'nauseous' smell so pervasive that it penetrates 'every house [...] and the clothes of the poor inmates sme[ll], and their food taste[s], of the damp grave!'.[66] The manager of Globe Lane cemetery aims to 'crowd the greatest possible quantity of corpses into the smallest space', and the is soil so replete with human relics that it is difficult to distinguish them from earth.[67] Smoke rises constantly from the chimney of the 'bone-house' on the cemetery ground, and its 'pest-bearing breath' ominously hints at the nature of the fuel, explaining how the burial ground owner makes room for new bodies.[68] Even making due allowance for the sensationalism of the genre, historical records suggest that this description did not stray much from that of actual burial grounds in London's poor neighbourhoods. Indeed, the fictional Globe Lane burial ground bears a remarkable similarity to the scandalous Spa-Fields cemetery.

The Spa Fields case broke early in 1845. On February 26, the *Times* reported that the residents of Clerkenwell brought before the magistrate the condition of severe overcrowding of the ground and the practices that were employed to make room for more bodies, which included disinterring the remains and burning them in the bone-house.[69] John Walters, the engine keeper of the parish, was repeatedly called to extinguish the fire of the bone-house. There, he found 'as many coffin boards as three men could carry, and a great deal of pitch' in the chimney, as well as a tell-tale smell of burnt decomposing flesh.[70] Residents complained about the dreadful smell of the burial ground, especially when the chimney of the bone-house released its fetid smoke, and voiced their concerns about the impact this had on their health. Also, they regularly spotted the gravediggers disinterring and mangling human remains, and then either mixing them again with the soil or carting them to the bone-house. Harriet Woods, a woman living next to the ground, witnessed the gravedigger jumping onto the coffins of her own children, and when on another occasion she 'called to him', he answered with 'a threatening language'.[71]

Although it had no impact on the legislation,[72] the Spa Fields case reached the attention of the authorities. The parish of the fictional Globe Lane cemetery, instead, ignores the requests for intervention coming from the dwellers of the adjoining houses. As for Mr. Jones, except when his rod taps a coffin that releases a particularly foul smell, he seems undisturbed by his working conditions and digs bare-armed into the sickly mud to disinter old, or not very old, coffins to make room for new ones.

Historical records suggest that real-life gravediggers were not so detached: to perform their 'disgusting' work, they often resorted to alcohol and sometimes sang to cheer themselves up.[73] This behaviour appeared insensitive in laypeople's eyes and raised the suspicion that their work made gravediggers indifferent to death, a conviction that cases such as Spa Fields reinforced. The fictional Mr. Jones's attitude towards the bodies he disinters reiterates this idea. The narrative indulges in graphic, gory details of the 'not entirely decomposed' bodies that the gravedigger breaks into pieces with his pickaxe.[74] The scene focuses on the 'upper part of a skull, with the long, dark hair of a woman still attached to it' which, it is suggested, was once 'a valuable ornament'.[75] The sentimental value attached to the hair contrasts with the 'coolness' with which Mr. Jones picks it up with the skull and throws it in the fire, where it 'hiss[es]' as it burns until 'the voracious flames li[ck] up the thin coat of blackened flesh which had still remained on the skull; and lastly devo[ur] the bone itself'.[76] Again, an unfeeling male character dedicated to the desecration of graves manhandles the female corpse, represented here by the hair, a symbol of female beauty. The dead body's femininity, connected to the Victorian idealization of the woman as a source of tenderness and good feelings in the man, underscores the gravedigger's lack of emotional response.

The extent of his indifference becomes apparent when he prepares his victuals. As he makes his coffee on the bone-house fire, the narrator notes that he keeps his water, his coffee-pot, and his coffee in the same 'foul place' where he burns the remains of the dead, and that he consumes his food in front of the same fire that he 'fed with human flesh and bones!'.[77] This ominously suggests that the cinders and ashes of the dead are contaminating his food, and therefore he is ingesting dead human bodies, which makes him a necrophage. Again, the historical source for this gruesome detail of Jones's professional routine can be traced to Spa-Fields: in *Second of a Series of Lectures* Walker writes in disgust that 'the watchman (?) of the [Spa-Fields burial] ground [...] was accustomed to eat his food in the *bone-house*, during the performance of his duties'.[78] Whereas he gradually 'became more fastidious', and ultimately ceased this habit,[79] Mr. Jones of Globe Town cemetery has no such scruples. On the contrary, there is an eerie connection between him and the personified fire, which, like a 'voracious' carnivorous beast, 'licked up' and then 'devoured' the female scalp and is 'fed' with human flesh and bones. By proximity, the gravedigger participates in the fire-beast's voracity by cooking on it and therefore, implicitly, feeding on the same fuel. This suggests that Mr. Jones's detachment originates in physical contamination, which hardened his soul. The constant contact with death, unmitigated by feelings of respect for the dead, dehumanized him.

The narrator stresses that Mr. Jones's activities are sacrilegious, bitterly observing that, while the bereaved console themselves with the idea that their loved ones are experiencing the '*tranquil slumber of the tomb*',

> the *last home* is invaded – the *quiet resting place* is rudely awakened with sacrilegious echo [...] and the corpse is snatched from the *tranquil slumber of the tomb* to be cast in the all-devouring furnace of the Bone-House. [80]

Expressions such as 'sacrilegious' and 'the corpse is snatched' appeared also in 'The Body-snatchers', connecting the two episodes and, consequently, the Resurrection Man and his gang on one side and Mr. Jones on the other. The narrator reinforces the connection, observing that the 'long flexible iron rod' Mr. Jones uses to test the ground before digging is 'similar to those which we have already described as being used by the body-snatchers'.[81] Still well into the 1840s, therefore, the popular mind associated the figure of the gravedigger with that of the resurrectionist and regarded it as the corrupt and corruptible party, another nuance that *Mysteries* reinforces. Indeed, while he is untouched by the decay that pervades his workplace, like Will Stephens in *Varney* Mr. Jones is rather sensitive to money. He usually enforces the rules of Globe Lane cemetery, which include the desecration of the tombs, because 'surely his superiors must know what was right and what was wrong!'.[82] However, he can make exceptions in case a better prospect presents itself, and Mr. Banks, the local undertaker, can offer him one. Upon request of the Resurrection Man, Mr. Banks has managed to access the body of the unknown stranger dead in the lodgings of the aged widow Mrs. Smith, and to introduce to her, the Resurrection Man disguised as the priest who will stage the fake funeral for the dead man. Now he only needs to bribe Mr. Jones to make sure Tidkins can snatch the body undisturbed. Jones appears determined not to transgress the rules of the burial ground, until Banks capitulates and offers to pay the gravedigger twice the sum he already takes.[83]

Mr. Jones's work ethic (or lack thereof) frames him as an agent of the power that ensures that the gap between the well-off and the pauper is protracted in death. He accepts the status quo, unless he can bend it to his own advantage. The proximity to decay that hardened him, and perhaps contaminated his physical body, also contaminated him from the moral and spiritual point of view. His character resembles that of the ghoul, the Arabic demon that becomes a cemetery-dweller, grave robber, and corpse-eater in English lore, and 'is described (by Edgar Allan Poe) as "neither man nor woman … neither brute nor human"'.[84] His presence in the narrative underscores the Londoners' blindness and anosmia about the state of urban cemeteries. Conversing with Mr. Banks, Mr. Jones remarks on the obliviousness of 'persons which dwells up in decent neighbourhoods […] and seems exceeding' proud of their fine houses and handsome shops' about 'the foul air that comes from places only hid by a low wall or a thin paling'.[85] The emphasis on the 'low' wall and 'thin' paling stresses the proximity between the dead and the living in Victorian London, of which the poor are constantly aware and the rich are blissfully, if not willingly, ignorant. Yet, Mr. Jones alleges, the West End cemeteries 'pursues just the wery same course as we does here'.[86] Indeed, the infamous Enon Chapel was decidedly towards the western side of the metropolis, and Mr. Banks cites it as an example.[87]

Furthermore, the ghoulish gravedigger is part of a systematic process of disrespect for the dead poor that was endemic to post-Act London—or England. The offensive state of intra-mural burial grounds combined with the assumption, sanctioned by the Anatomy Act, that the poor's right to a decent interment was inferior to that of middle- and upper class people, tacitly authorizes the fictional Jones to desecrate the tombs in his ground and to take bribes to allow others to do the same.

The connection between Mr. Jones and the Resurrection Man is significant from this perspective: both connive, both perform similar activities, and, most notably, as with the other monsters examined so far, Jones also disrupts the boundaries between life and death bringing dead matter back to the world of the living. Consequently, he is in the perfect position to 'resurrect' the practice of bodysnatching. Thirteen years after the passing of the Anatomy Act, the figure of the gravedigger could still be marketed as a sacrilegious and untrustworthy party involved in the black market for cadavers. After all, was the medical fraternity not demanding yet more bodies?

The three monsters the bodysnatching episodes of *Mysteries* present originate in the connection between displaced bodies and anatomy studies that existed in the popular mind since the beginning of the century. In the narrative, the triad builds a sensational exposé of social injustice, as each of its elements contributes to extend the gap between the rich and the poor to the disposal of the remains after death. Mephistopheles, Faustus, and the Ghoul cater to the fears of the series' working- and lower middle-class audience about the ignominy of the anatomist's scalpel, curtailing the distance between the pre-Anatomy Act era and the audience's present time through connections with such recent scandals as Enon Chapel and Spa Fields. Unlike in *Manuscripts* and *Varney*, anatomy has no redeeming features in *Mysteries*: there are no heroic doctors, bodysnatching has no higher purpose than that of satisfying a greedy, voyeuristic surgeon, and is performed by monstrous individuals.

Such a grim picture contributes to *Mysteries*' larger teleological purpose, which is possibly its most prominent feature: notwithstanding its use of sensation and melodrama, this is a narrative with a political agenda.

2 Poverty and Wealth, Language and Power: The Anatomy Act and Political Discourse

The narrative's political undertones, which constitute the core of the series from the opening prologue, strongly influence medical discourses in *Mysteries*. The text calls the reader's attention on the social gap between the rich and the poor, which by the 1840s had become the trademark of Victorian society, and on its most famous product, the metropolis of London:

> There are but two words known in the moral alphabet of this great city; for all virtues are summed up in the one, and all vice in the other: and those words are
>
> WEALTH. | POVERTY.[88]

This opening concept becomes the paradigm of the whole work. A line visually separates the words 'wealth' and 'poverty' in the body of the text, while the full stops signify their finality. Stephen J. Carver observes that the juxtaposition between social classes in Reynolds' series constitutes 'the recognizable duality present in all urban writing', while 'the additional language of class war' singles out *Mysteries* in this subgenre.[89] In contrast with the penny bloods' tendency, noted by Crone, to reconcile the reader with the status quo,[90] *Mysteries* sent its readership a revolutionary message, which is compatible with the agenda of all Chartist fictions. Haywood explains that, maybe more than anything else, the Chartist movement found purpose in the fight against the 1832 Reform Bill and its 'treacherous' disempowerment of the working class.[91] Consequently, he contends, Chartist fiction can rightfully be defined 'propaganda' as it was 'more aggressive' in proposing a solution to the socio-political prevarications than other contemporary forms.[92] Chartists were aware of the 'ideological role of literature': to them, 'art was an expression of the political condition, not its antithesis or its imaginary solution'.[93] From this perspective, then, the first series of *Mysteries* can certainly be considered Chartist fiction, as it expressed explicitly, perhaps more than any other penny blood, the injustice inherent in the current socio-political situation.

The series' sensationalism somewhat undermined its political intentions in the eyes of the educated reader, both coeval to the series and future, and Reynolds's use of melodrama indeed cost him his reputation for a long time.[94] However, Haywood notes that the use of melodrama, instead of impairing the political value of the narrative's radical message, promoted its reaching a wider, more diverse audience than other forms of fiction.[95] Significantly, this message included discourses related to anatomy and dissection. From a medical-historical perspective, the discourse on social injustice can be broken down into two different sub-discourses. Firstly, as in *Sweeney Todd* and *Varney*, in *Mysteries*' brutal society knowledge is power. Secondly, although the series does not mention explicitly the Anatomy Act, it questions its effectiveness in enforcing legality and guaranteeing the safety of the public.

The chief trait that distinguishes the Resurrection Man from Sweeney Todd and Varney is that, while the vampire and the barber endeavour to conceal the truth about themselves, Anthony Tidkins is very open about himself, and often about other people, especially if this grants him a position of power. This difference originates in the fact that Tidkins hardly ever feels threatened by truth, while he is the one who threatens other characters with the disclosure of dangerous truths. Unlike Todd, Tidkins never bluffs when he asserts that he might ruin a character with a piece of information he possesses. When he is in danger of being prosecuted, he does not hesitate to turn in his accomplices and, after Richard antagonizes him, he taints his reputation revealing his past as a Newgate convict more than once. The Resurrection Man can be read as a piece of walking truth, the ugly truth that cannot be silenced, and this is not limited to his revealing other characters' secrets: as a product of the social inequality that regulates the city, he embodies the uncomfortable truth of the injustice underpinning the Victorian metropolis itself, as Reynolds represents it.

Since London will not acknowledge Anthony Tidkins' uncomfortable existence, he makes his presence undeniable by becoming a ubiquitous, dangerously active entity in society, one that transcends social boundaries. Taking this point a step further, I would state that the threat Tidkins poses to middle-class characters symbolizes the possibility that, though the current socio-political situation primarily harms the poor, its consequences may befall the middle class as well. The Resurrection Man's profession is as obsolete as the exclusion of the lower strata from government and the legislation that subjects them to the anatomist's scalpel. Suggestive as he is of the dark times of resurrectionism and burking, Anthony Tidkins signifies the insubstantiality of the idea that the Anatomy Act can solve the anatomists' demand for bodies and the threat it poses to individuals regardless of their station in life.

Haywood's analysis of adverts in Reynolds's publications shows that he was aware of the variety in the composition of his audience, in terms of both social extraction and gender.[96] Haywood contends that, for 'respectable' Victorians, the most damning feature of the cheap serialized fiction Lloyd and Reynolds produced was that its appeal 'transgressed conventional [...] boundaries'.[97] Therefore, *Mysteries*, and the penny blood genre more broadly, was undesirable because it was the tangible proof of the permeability of social barriers, which was proven to be just as thin as the vertical line between the words 'wealth' and 'poverty' that appeared in the 'Prologue'. This subversive concept underpinned *Mysteries*' plot itself, suggesting that the boundary that excluded the working class from view and from power would not shield the upper and middle classes from the consequences of this exclusion, which would find their way through the social barrier in the same way the lurid *Mysteries* series did.

The narrative conveys discourses about the danger of social inequality by physically and socially moving the characters between the two apparently separated states of wealth and poverty. Haywood explains the tendency of authors such as Reynolds to 'degrad[e] heroes and heroines' as an attempt to 'personify' the 'frustrated respectability' of the working class.[98] Ellen Monroe and her father, but also Richard and, to an extent, Eliza, are instances of this narrative strategy. The movement of an individual from wealth to poverty was, after all, an event still in the realm of possibility. The actual *frisson*, however, is constituted by the Resurrection Man and his gang's eerie ability to break socio-geographical boundaries, reaching out from the world of poverty and crime and threatening the world of wealth. Tidkins' transgression of socio-geographical boundaries reflects the genre's transgression of social and cultural norms. He and his gang constantly infringe the social conventions that forbade their trespassing in the space of respectability and wealth. The first tomb they open in the narrative is that of a middle-class girl; they almost kill Richard Markham and Eliza Sidney, two upper middle-class characters; and they go as far as attempting to break into Buckingham Palace, which to them is 'no more than another crib to crack'.[99] No place is too high for Anthony Tidkins, not even the very heart of the British Empire, and neither the property, nor the secrets of his social betters are beyond his reach.

Mysteries is a penny blood with a resurrectionist villain, and it is therefore natural that the threat posed by the permeability of the illusory boundary should concretize in physical destruction, more precisely annihilation through dismemberment. The narrative directly connects this danger to the lack of communication between the two worlds, particularly to the middle-class characters' ignorance of the language of poverty. From several episodes, it emerges that characters from the world of poverty react to the threatening middle- and upper class power by creating their own code of communication. As in *Sweeney Todd*, this separate code grants its speakers the power of excluding others from discourse. Richard and Ellen ignore the code of poverty and suffer the consequences of their ignorance; furthermore, the narrative seems to suggest that, to an extent, their ignorance is a shortcoming on their part.

Richard, honourable to a fault, refuses to assume any honourable intent in Tidkins when he meets him near Count Alteroni's villa, and treats him as his social inferior. He knows nothing of Anthony except that he is a former Newgate convict, which condition they share and makes their positions in society not very different after all. This is something Richard refuses to acknowledge. Therefore, when Anthony asks him what he is doing in that neighbourhood and who was the girl with whom he was talking, he is outraged: 'surveying the ruffian with mingled indignation and disgust', he exclaims 'And by what right do you dare to put those insolent queries to me?'.[100] His gaze averted, the Resurrection Man 'coolly' replies that, in that case, '[he] can precious soon ascertain all the truth for [him]self'.[101] Far from catching the threat implicit in these words, Markham displays his middle-class outrage at being treated as a peer by a former Newgate convict. As Anthony turns towards the house of the Alteronis, he cries:

> Wretch! What do you mean to do? [...] you do not know that that abode is sacred – that it is the residence of probity, innocence and honour – that if you were to breathe a hint who [sic] and what you are, you would be spurned from the door?[102]

By defining Isabella's house as a 'sacred' place where 'probity, innocence and honour' reside, Richard is implying Tidkins' foreignness to these virtues, which makes him unwelcome among *respectable* people. He marginalizes Anthony, reminding him of his low, *disreputable* social station. This is a mistake. Since Richard connects power to social standing, he fails to see that Anthony's knowledge gives him power, and that his position is far from subdued. Tidkins bitterly replies that he is 'accustomed to *that* in this Christian land', and then adds that he may just

> ask for alms [...] at that house, and thereupon state that the gentleman who was just now walking with the young lady [...] was a companion of mine in Newgate – a communication which will tend to preserve the innocence, honour, probity and all the rest of it, of that family.[103]

To Richard's protests that he cannot be 'so base' as to 'betray' him and 'ruin his reputation', Anthony retorts: 'And why should I have any regard for you, since you receive and treat me as if I was a dog?'.[104] Of course, Tidkins blackmails Richard to keep silent with the Alteronis about Newgate, asking for a sum that would leave the young man ruined. 'Why do you persecute me in this way?' Richard asks, 'why should you seek […] to annihilate all my hopes of again establishing myself in an honourable position in society?',[105] at which Tidkins ferociously replies: 'My law is the law of the world – *the oppression of the weak by the strong*; and my right is also that universal practice – *the right of him who takes what will not dare to be refused*'.[106]

This exchange marks the reversal of roles between Markham and the Resurrection Man, and the moment in which Anthony Tidkins becomes Richard Markham's nemesis. Richard's alienation from the world of Anthony Tidkins made him unable to communicate with a lower class individual. His rhetorical speech on virtue and rights and his emphasis on his efforts to regain a 'honourable', that is, *respectable*, position in society antagonize Tidkins, who has been excluded from the right to respectability by the class to which Richard Markham belongs. His own upbringing led him to perceive the supposed moral superiority and justice of the middle class as an oppressive power, which he resents and transforms into his own distorted code of oppression. Tidkins' awareness of the power of knowledge has obvious Chartist undertones, since it emphasizes the importance of knowledge to the oppressed to overcome the oppressor. He thus becomes a subversive entity in the narrative: he is the powerless individual who overpowers his social betters using knowledge as a weapon.

Markham's ignorance of this type of knowledge, and of the code of communication of the world of poverty, almost costs him his life on other occasions. Since poverty and its knowledge are the key to understanding not only the language, but also the space of poverty, he gets lost in the slums where the Resurrection Man lures him, and repeatedly falls in his traps. The first time, Markham is imprisoned in the Resurrection Man's house, where he is almost killed and sold for dissection. The second time, Tidkins, posing as his long-lost brother Eugene, writes a message asking Richard to keep an appointment at a place ominously called Twig Folly. Although the message sounds strange, Richard reflects: 'Who could wish to injure *me*? Who would wish to take *my* life? […] The idea is preposterous!'.[107] On the next page, Richard is falling into the river, pushed there by the Resurrection Man, and he escapes death again, thanks to the help of his friend Ellen Monroe in male disguise.

Similarly to Richard, Ellen ignores the code of poverty in consequence of her secluded middle-class life. The only daughter of 'fond, but too indulgent parents' who raised her 'in the lap of luxury',[108] when she falls into disgrace and becomes a destitute seamstress in Golden Lane, she cannot communicate with the dwellers and does not understand who she can or cannot trust, which makes her the target of the 'old hag' who lives in the same building. When she first approaches Ellen, the narrative underscores the failure in the communication between the two women. To Ellen, her neighbour's words are 'a strain' that she can 'scarcely comprehen[d]'; although she understands she is being offered 'a more pleasant and profitable mode of earning money', she instinctively recoils and 'tremble[s] at the words that issu[e] from the crone's mouth'.[109]

The scene frames Ellen as a foreigner in the environment of the slum, unfamiliar with its language and costumes. Notably, the narrator defines the failure in communication, and Ellen's status of foreignness, on the basis of moral parameters: '[t]he words of the old woman were [...] unintelligible to [her]' because '[t]he soul of Ellen was purity itself'.[110] In their second meeting, Ellen fails again to understand the woman, and listens 'as if she [is] hearing a strange language which she [is] endeavouring to make out'.[111] This passage explicitly compares the two estranged classes to two foreign cultures characterized by different linguistic and moral codes. When Ellen finally understands (although it is not stated explicitly) that she is being offered to become a prostitute, she reacts according to the code of Victorian middle-class female respectability, covering 'her blushing cheeks with her snow-white hands' and bursting into tears.[112] In the world of poverty, though, the middle-class code does not work, and her innocence becomes naiveté, exposing her to the old woman's plans. To strengthen her grasp on the girl, the old woman finds her a series of temporary occupations among artists—a statuary, a sculptor, a painter, a photographer, and an illusionist. In all these works, Ellen is in contact with men unchaperoned and, on most occasions, she must pose naked for them. The instalment, appropriately named 'The Road to Ruin',[113] shows Ellen selling herself piecemeal, corrupting her soul in the process. Even the narrator, who was praising her virtue only a few pages earlier, coldly concludes:

> Suffice it to say that [...] [a] tainted soul now resided in a pure body. Every remaining sentiment of decency and delicacy was crushed [...] Pure souls have frequently resided in tainted bodies [...] but here was essentially a foul soul in a chaste and virgin form.[114]

However, Ellen is only in part responsible for her state, according to the narrator: her ruin is to be ascribed to

> dire necessity – that necessity which became an instrument in the old hag's hands to model the young maiden to her purpose. [...] The wretch knew the world well, and was able to calculate the influence of exterior circumstances upon the mind and the passions.[115]

Ellen's transformation from 'solitary lily in the midst of a black morass swarming with reptiles'[116] into prostitute contributes to the narrative's discourse on the power of knowledge. Whereas Ellen's secluded, pampering life made her dangerously ignorant of the world and unprepared for a social fall, the old woman from the slums 'knows the world well'. The hostile environment where she lives hardened her nature and made her knowledgeable enough to pose a threat to heedless strangers.

Characters such as Richard and Ellen attacked the foundations of a middle-class reader's sense of safety. They conveyed the subversive message that the power of social standing is an ineffective defence against poverty, both as a condition in which individuals may fall and as a source of threats for the unaware. This message underpins also the narrative's anti-Anatomy Act discourse.

As the juxtaposition of the pauper's knowledge to middle-class character's ignorance suggested that the boundary between wealth and poverty was permeable, so did the graphic scenes of misery, corruption, and filth that constituted the narrative's anti-Anatomy Act discourse. These scenes suggested that destructive effects of the Act on the poor were not as safely distant as the middle-class imagined, and reiterated practices from the dark days of anatomy studies that could easily cross social boundaries. Richardson notes that the Anatomy Act failed to produce a change from the era of burking and bodysnatching,[117] a point that emerges forcefully in the discourse about the injustice of the current legislation on the disposal of the pauper's remains *Mysteries* constructs. The episodes relating the theft of the body of the unknown man in Globe Lane are key to understanding this discourse, as they revolve around the set of characters that in real life propelled, more or less overtly, the mechanism of the Anatomy Act: an unknown body; a poor woman; an undertaker and a gravedigger; and a bodysnatcher.

The old man who dies in Mrs. Smith's lodgings had been ill for a while, and the widow nursed him, helped by the wife of the Buffer, one of Anthony Tidkins' accomplices. As soon as he dies, the Buffer's wife immediately reports the death to her husband and the Resurrection Man, together with the news that the surgeon who examined the body let her understand that he would pay her husband generously for disinterring it. The Resurrection Man immediately contacts Mr. Banks, the undertaker of Globe Town, who sets off to Mrs. Smith's house. The widow does not suspect she is letting in an undertaker: the solemnity with which Mr. Banks declares that 'the least we can do is to show a feeling of weneration [sic] for our deceased friends by consigning them in a decent manner to the grave' induces her to think that he is a relative, or a minister.[118] She is relieved that someone is there to 'superintend the funeral', because 'it's a great tax on a poor lone body like [her] to have such an undertaking to attend to', and Mr. Banks slyly reassures her: 'I'll *undertake* the undertaking'.[119] His attitude convinces Mrs. Smith that he must have 'some legitimate authority for his present proceeding', and she shows him the body. Finally, when she understands who she let in her house, her reaction—'"Undertaker!" ejaculated the widow'[120]—is not too different from Richard Markham's surprise at hearing Anthony Tidkins' profession at Newgate.

Indeed, neither Tidkins nor Banks is suspect until he announces his profession, until they name themselves. Undertakers were deemed to be rather untrustworthy, and the *Rex versus Feist* case in 1858 would confirm these suspicions. The woman's reaction challenges the undertaker's legitimate authority: he is neither a relative, nor any form of parish or religious authority, so he has no right to access the body. Still, the woman is happy to be guided, and the fact that she refers to herself as 'a poor lone body' is meaningful. The expression, typically Scottish and probably absorbed in the English language, is generally a substitute for the pronoun 'oneself' and may entail 'sympathy'.[121] In this case, the expression suggests that she is akin to the unknown dead man in her lodgings: being a pauper and a widow, she is a very likely candidate for the surgeon's slab, especially if she ends up in the workhouse.

Having gained the widow's trust, Mr. Banks invites a minister of his choice to perform the funerary rites, and this is, of course, the Resurrection Man in disguise.[122] In the course of his visit to Mrs. Smith's house, the widow asks him whether he thinks 'there's such people as resurrection men now-a-days?'; Tidkins brazenly replies that 'society has got rid of those abominations' and now surgeons get their bodies '[f]rom the hulks, the prisons, and the workhouses'.[123] Mrs. Smith 'revolts' at the idea, and exclaims: 'Well for my part [...] I think it's wery hard if, after paying rates and taxes for a many – many year, I should be obleeged to go to the workus, and then be cut up in a surgeon's slaughter-house at last'.[124] The widow did not know that the workhouses were currently a source of bodies for anatomy studies, and her distress upon receiving the information is obvious, because she knows that a person like her is likely to enter the workhouse. Her first reaction is of indignation: she is, after all, a ratepayer who, in her poverty, always paid her 'rates and taxes'.

This remark undermines one of the founding principles of the Anatomy Act, which considered the use of the pauper's bodies for dissection as a payment for their dependence on parish (that is, on the ratepayers') support in life. Besides the obvious lack of sympathy, the fallacy of this principle was the assumption that all the paupers who ended up in the workhouse had always been a burden on the parish, whereas records prove that this was not always the case. A family might fall into poverty after the death, illness, or injury of one or more of the people who provided the income, or perhaps illness might force a pauper to seek medical care at the workhouse infirmary. Such was the case for Charles Greenland, one of the people whose remains were wrongfully dissected in the *Rex versus Feist* case. Illness had made Charles unemployed, and his family decided that 'their best option' would be for him to receive medical care in the workhouse infirmary,[125] where he died. Charles's remains were sent to Guy's hospital for dissection while the unsuspecting family buried a stranger. Before falling ill, therefore, Charles did not depend on the parish, and yet, he became a subject for dissection at Guy's. Although preceding *Rex versus Feist* by almost a decade, the Resurrection Man's dialogue with Mrs. Smith questions both the solution the Act provides and the people who demanded this solution, namely, the anatomists.

Mrs. Smith's horror at the idea that workhouse inmates should end up being 'cut up in a surgeon's slaughterhouse' indicates that the image of the butcher-surgeon was still current in the late 1840s. *Mysteries* exploited the sensationalistic undertones of this image to convey the idea of an inhumane medical fraternity still engaged in butchering the poor and powerless. The tone of this representation was set during Mr. Banks' first visit, when Mrs. Smith wonders at his kindness:

> I always heerd [sic] say that butchers and undertakers was the most unfeelingest of men. They never let butchers set on juries; but I'm sure if undertakers is so milk-hearted, *they* may set on juries.[126]

Mrs. Smith's opinion on butchers influences the subsequent comparison of dissection rooms to slaughterhouses and, indirectly, of anatomists to butchers, including surgeons among the 'most unfeelingest of men'. This image suggests that the current legislation enforces the role of the surgeon as butcher by subjecting the most exposed individuals in society to the further humiliation of dissection, and in so doing, wronging honest citizens.

The political discourse on the disposal of paupers' remains maintains the focus on the unfairness of the law with the introduction of the grave-digger Mr. Jones in the narrative. The management of Globe Lane burial ground deprives the people who are interred there of a dignified conclusion to their life, and yet, the parochial authorities dismiss the complaints of the residents and tell them to 'prefer an indictment at the session'.[127] The narrator grimly remarks that with this move, the authorities prevent the poor from being able to take action:

> Such a process is only accessible to those in possession of ample means; for the legislature has purposefully rendered law, – that is the power of obtaining justice [...] – a luxury attainable only by money. [...] [W]ho ever thought of legislating for the poor? Legislate *against* them, and it is all well and good: heap statute upon statute [...] encumber the most simple form with the most intricate technicalities – diversify reading and expand in verbiage until the sense becomes unintelligible – convert the whole legal scheme into a cunning web, so that the poor man cannot walk three steps without entangling his foot in one of those meshes of whose very existence he was previously unaware, and whose nature he cannot comprehend [...] do all this and you are a wise and sound statesman; for this is legislating *against* the poor – and who, we repeat, would ever think of legislating *for* them?[128]

With this comment, the narrator illustrates the procedures of exclusion of the poor from the law discourse: they cannot speak the language of law, which is the language of wealth, where education grants access either to the code itself, or to people who can interpret it. In the same way as Richard and Ellen cannot survive in the slums, the poor cannot survive in this world, and become the hapless prey of the system.

Passages such as this are typical of Reynolds's own radical propaganda incursions in his fiction. Humpherys observes that Reynolds's 'strident, imperative, and frequently hyperbolic narrative voice' could sound remarkably like Reynolds's own tone, as he used it in his own

Sunday newspaper *Reynolds's Weekly*.[129] This technique created a continuity between the literature of information and the literature of escapism he provided for his audience, and contributed to building what Humpherys terms the 'exposé of institutions'[130] that constituted Reynolds's purpose in writing *Mysteries*. His aim, Humpherys argues, was that of 'educating generally lower class readers about the depredations made upon them'.[131] Haywood also notes Reynolds's use of the narrative voice to intrude in the narrative to 'demystify the action, and to remind the reader that all plots are socially and politically generated'.[132] Melodrama, Haywood continues, facilitated this process as it enabled the narrative voice to bring the political and social affairs of the present time to the story, exploiting 'what Peter Brooks calls melodrama's 'hermeneutic excess''.[133] Reading the passage above from the perspective of the disposal of the remains of the poor, it can be stated that the narrative-propagandistic strategy exposes the ambiguous language used in the Anatomy Act, which left plenty of room for interpretation and made it 'the bureaucrat's bad dream'.[134] The narrative thus extends the paradigm of reciprocal foreignness of language and knowledge between classes that the social gap creates to the discourse around anatomy, the poor, and the law. The passage illustrates the procedures of inclusion and exclusion from said discourse, highlighting the lack of balance in power the Anatomy Act creates, contributing to maintain the working class in a subject position.

Mysteries represented law as twisted and unintelligible, and made so on purpose by a system that had no interest in protecting the working class. The right to justice does not belong to the pauper, because its discourse and language belong to the class in charge. 'The Gravedigger' episode resonates with the echoes of the inability of the poor, observed by Hurren,[135] to understand the ambiguous text, and therefore the full meaning, of the Anatomy Act. The emphasis on the complications of the legal language—'***encumber*** the most simple form with the ***most intricate*** technicalities – ***diversify reading*** and ***expand in verbiage*** until the sense becomes ***unintelligible***'[136]—is remarkably compatible with the effects of the Act's text on the semi-literate reader. Law and its unintelligible language are represented as a weapon in the hands of the middle and upper classes 'against' the poor, used to hide from them knowledge of their rights regarding the disposal of their own remains. Its language is the language of literacy and education—the language of power.

The Resurrection Man is the only character in the narrative who escapes the exclusion from discourse that regulates the power dynamics between the rich and the poor. He avoids the unintelligibility trap, thanks to his extremely plain, intelligible speech, which testifies to his alarming ability to move through the vertical layers of Victorian society, deploying all his resources to gather information that he can later use to his own advantage. The Resurrection Man and his subversive reverse justice code threaten middle-class Richard Markham, and the graves of different people in the narrative, regardless of their social station in life, because he does not respond to the language of law, nor is he subject to it. Anthony Tidkins and his boundary-breaking language represent the terrible possibility that the unfairness of a socio-political organization that regulates the life of individuals to the grave and beyond according to wealth and poverty might rebound on the unsuspecting middle class.

Mysteries shows the lower and the middle class engaged in using language to vie for supremacy: as the poor could not understand legal language, so the rich could barely understand the language of the poor classes. Outside fiction, however, in the arena of government and legislation, the poor lost the battle. Their power was confined to the slums. Conforming to this dynamic, *Mysteries* systematically represents the poor as voiceless, unintelligible, and the legislation regulating the disposal of their remains as written *against* them. Representing this lack of balance in power and its connection to language, the narrative attempts to make readers aware of their position with respect to it, while at the same time translating the unintelligible language of law for them. The geographical movements that accompany the characters' experience of different social contexts constitute the spatial scaffolding of the discourse on power, knowledge, and dissection in the narrative. In *Mysteries*' labyrinthine London, space speaks of the anxieties about dismemberment and annihilation that haunted the geospace of the Victorian city.

3 The Fall and the Labyrinth: Mapping the Way to Truth

Similarly to *Sweeney Todd*, *Mysteries* strives to create a perfect superimposition of the Victorian metropolis in the narrative to the geospace of Victorian London. This process includes the construction of the urban landscape on an 'imaginative diagram' of 'oppositions' to expose social contrasts, which Humpherys notes was a strategy used also by several authors of fiction, most notably Dickens and Pierce Egan, and by 'political cartoonists'.[137] In so doing, Reynolds endeavoured to represent his readers' space with its connections to the different aspects of their life, including their anxieties and frustrations. In the narrative space of *Mysteries,* we encounter locations that are directly connected with the history of anatomy in the metropolis, through which the narrative explores the role of medicine in enforcing social inequality. These spaces become a highly sensational, graphic landscape in which dramatic scenes of poverty and crime unfold. A representative instance is the use of Bethnal Green, the area where the historical London burkers lived, as the abode of the Resurrection man. The powerful historical charge of spaces such as Bethnal Green contributed to constructing the narrative's urban space as a reflection of the unbalanced power distribution in the metropolis, and this representation could not but focus on the East End area. Within this narrative space, Reynolds rewrote the tropes of the trapdoor and the labyrinth, which, as we have seen in *Sweeney Todd*, had become part of metropolitan culture and tapped into anxieties related to the city's destructive potential. While they contributed to discussing the social chasm in the story, these tropes also exploited and reinforced popular discourses around death, crime, and medicine in the space of the London East End. The narrative space of *Mysteries* thus became a social-geographical commentary inscribed within an exciting narrative, which was meant to create in the reader the illusion of living into their narrative space.

The East End was synonymous with poverty, unsanitary housing, and precarious health; logically, therefore, the Anatomy Act had a strong impact on its inhabitants. Notably, it was also the space of hospitals: south, the London Hospital in Whitechapel; west, St. Bartholomew's in Smithfield; north, St. Luke's Lunatic Asylum, adjacent to the slum of Old Nichol, and the private anatomy schools of Little Windmill, next to Bethnal Green; and, further south, beyond the river, St. Thomas and Guy's Hospital, located in the analogous area of Southwark. The evolution of the geography of the city caused the hospitals to become part of the degraded scenery of the Victorian metropolis because, in fact, they were originally built to serve the poor: St. Bartholomew's and the St. Thomas's and Guy's were founded to provide health care for the pauper, and St. Luke hosted specifically the lunatic poor. The London Hospital in Whitechapel, built in 1740 under the name of London Infirmary and moved to Whitechapel in 1752, was meant to treat 'the sick poor among "the merchant seaman and manufacturing classes": the east End community of the time'.[138] This geospace of medicine and poverty became the space of ***murder*** connected to medicine and poverty when the 'Italian Boy' case exploded. It is therefore unsurprising that this area should become in the narrative a space in which the crimes that the Anatomy Act should have dispelled are instead uncannily alive.

Anthony Tidkins' presence pervades the narrative space of the East End. Indeed, Hackenberg suggests that the reader, as the characters in the story, is threatened with encountering him at every 'turn of the page'.[139] Although his address varies slightly across the narrative, the fictional Tidkins lives in an area that corresponds to the geospatial Bethnal Green where Bishop and Williams lived. The house in which he imprisons Richard Markham is near Birdcage Walk, which merges with Crabtree Row, bordering the Nova Scotia Gardens, where the London burkers lived. Later, Tidkins himself will give his own address as 'the Cambridge Road corner of Bethnal Green Road'.[140] Bethnal Green Road is south, and slightly further east, of Birdcage Walk, but it is still, unmistakeably, Bethnal Green, East London. Shoreditch, where he snatches the corpse of the girl, is immediately west of Bethnal Green, and it is also the site of Holywell Mount cemetery, which, as we have seen, had record of being a bodysnatching site. Fanny Pigburn, one of Bishop and Williams's victims, was also from Shoreditch. Finally, Globe Town, where Tidkins resurrects the body of Mrs. Smith's late tenant, borders Bethnal Green and Whitechapel, where the London Hospital was rebuilt in 1752 (Fig. 1).

The systematic superimposition of the narrative space of the Resurrection Man on the geospace of burking and bodysnatching makes the narrative resonate with the echoes of a newspaper report. Furthermore, as Haywood observes,[141] and as shown in the previous sections, Reynolds tended to include in his fiction a considerable amount of non-fictional material. The extensive discussion of Enon Chapel between Mr. Jones and Mr. Banks in 'The Grave-digger' is a perfect example of this operation. The Enon Chapel scandal was discussed in the newspapers as late as 1847 and in 1842, the *Times* dedicated a series of articles to the case, which included passages from George Walker's report *The Grave Yards of London*.

In *Grave Yards*, Walker wrote that in Enon Chapel, during the summer months, an insect 'which had the appearance of a common bug

Fig. 1 Greenwood, Christopher and John. Map of London, from actual survey, comprehending the various improvements to 1851. Detail of Bethnal Green © British Library Board 03/07/2018. Shelfmark: Maps Crace Port. 7.265. Item number: 265. The yellow circle marks the area of the Nova Scotia Gardens, with Crabtree Row in red below it and Birdcage Walk in green on the right

with wings' was observed 'crawl and fly' through the building.[142] It was assumed to come from the festering bodies that were piled up to reach the thin floor board, and the children of the Sunday school had named it 'body bug'.[143] In *Mysteries*, Mr. Jones and Mr. Banks describe the appalling conditions of the chapel's vault with great accuracy, down to the shape of the 'body bugs', an insect 'just like the common bug, and with wings' produced by the decomposing bodies under the chapel.[144] Furthermore, Walker wrote in *Grave Yards* that 'second-hand coffin furniture, (nails, more especially) may be found by the hundredth weight, at many of the 'dealers in marine stores'' and that '*coffin wood has been extensively used as an ordinary fuel in low neighbourhoods*'[145]; the *Times* reported this passage verbatim in October 1841.[146] In *Mysteries*, Mr. Jones boasts about the 'sight of coffin nails [he] sells every month to the marine-store dealers', and claims to feel like a fool for not selling 'the coffin-wood for fuel, as most other grave-diggers does'.[147] The inclusion of non-fiction material, combined with the superimposition of the narrative space on the geospace, gives the impression that the events narrated are happening *as* the series is being published. Reynolds embedded his fiction so deeply in his reader's reality that the barrier between the two thinned. Consequently, such passages as the conversation between the gravedigger and the undertaker could raise the readers' indignation. Likewise, the crimes of the Resurrection Man, in theory suppressed by the Anatomy Act, could be imagined to be happening in the readers' reality.

The scenes in which Anthony Tidkins performs acts of burking and bodysnatching are sensational and crude, but they always include details that anchor them to the intended reader's reality, such as the uncanny similarity between his modus operandi and that of Bishop and Williams. Such a degree of realism made it easier to imagine that Tidkins was perpetrating *at the moment* crimes that theoretically belonged to the barbarous past of the London burkers. The narrative represents doctors requiring his services, as the actual medical fraternity did when they purchased murdered and stolen bodies coming from the same locations the narrative described. The annihilation of the space–time distance between readers and events narrated ominously suggested that in the geospace equivalent to the narrative, the exploitation of the dead poor was still going on. This operation constructed both the fictional and geospatial East End as the locus of danger that threatened the integrity of the body, a liminal space that rejected legality and was intrinsically deadly. These features emerge both in the labyrinthine design of this narrative space and in the movement the characters perform through it.

Mysteries exploits the trapdoor trope explicitly connoting the fall as both geographical and social. While from the spatial point of view it represents a sudden drop from the world of life above to the world of death below, its association with murderous/criminal characters also defines it as a movement from respectability to social ruin. The fall is also the chief movement of the narrative and it opens the series: in Chapter 1, Eliza Sidney (disguised as Walter) is lost in the terrifying neighbourhood of Smithfield, is caught in a storm, and seeks shelter in an ominous-looking old house. After a rather gingerly exploration, she finds that in one of the rooms, 'in the middle of the floor—only three feet from the spot where he [stands]—there [is] a large square of jet blackness', and immediately '[a]n indescribable sensation of fear [creeps] over him'.[148] Discussing images of water underground in narratives as inspiring terror, Bachelard notes that the 'shiver' that the character 'sense[s]' before this subterranean water goes beyond humanity to become 'an anthropo-cosmic fear that echoes the great legend of man cast back into primitive situations'.[149] It is a feeling that speaks to a more primeval level of our mind, and indeed Eliza perceives, feels, rather than sees, the deadliness of the plunge into darkness the black square represents, similarly to the protagonist of Poe's tale 'The Pit and the Pendulum'. Poe's tale, pre-dating *Mysteries* by two years, likewise shows a panic-stricken protagonist in a dark, underground space, on the verge of a dark 'pit' with water at the bottom. While Poe's protagonist avoids the jump into darkness, by the end of episode three (the title of which, 'The Trap-door', clarifies the nature of the black square on the floor), two villains push Eliza into that 'ominous blackness' which is 'the mouth of a yawning gulf' on the Fleet Ditch.[150] Before falling, Eliza cries: 'Do not—*do not murder me!*',[151] confirming that the plunge in the trapdoor brings death. Only three instalments later, the reader learns that she survived the fall, and that her miraculous escape led her to a tour of the slums, inclusive of a visit to a rookery, during which she learns about the miserable lives of its inhabitants.

Richard Markham also experiences, and survives, the fall in the Resurrection Man's lair, where Tidkins left him for dead with his terrible Mummy. Taking the woman by surprise, Markham tries to learn from her the name of the unknown corpse he saw laying in the kitchen, in the hope of alerting the dead man's friends. The Mummy tricks Richard into entering a dark room, where a trapdoor gives away under his feet. In the next chapter, though, the readers learn from a furious Tidkins that the Mummy plunged Markham, not towards death, but towards safety: 'Did I not tell you a month or so ago that the wall between the hole and the saw-pit in the empty house next door has given away!'.[152] Richard escapes and reaches a station house, where he asks for help. As with Mark Ingestrie in *Sweeney Todd*, Eliza and Richard emerge from the plunge in the underground bringing back truth about an unseen reality of misery and death. However, Richard's account of the horrors of Birdcage Walk yields hardly any results, because he cannot guide the police back to Anthony Tidkins' house. After his escape, he runs at break-neck speed for one hour, covering 'many miles of ground',[153] the prolonged effort of the run conveyed through the three-time repeated phrase 'He ran – he ran'.[154] In this mighty effort, he pursues 'tortuous paths and circuitous routes',[155] and emerges in the district of Wapping, south of Whitechapel, in the docks area east of the Tower of London. From there, he cannot trace his way back. The downwards movement of the fall in the East End thus appears to be closely related to the unmappable tortuous space of the slum, which Richard and Eliza experience both before and after the (supposedly) deadly plunge.

Mysteries middle-class characters' ignorance of the language to communicate in the slum is matched by their lacking the geographical map to navigate its labyrinth. The labyrinth is the intrinsic nature of the East End: being synonymous with the unknowable, the concept expressed the dangers (imagined and real) and lack of logic that defined the East End geospace. The middle-class character is lost in the twisted streets of the narrative labyrinth, as helpless and as exposed to fatal danger as a Gothic heroine. Eliza/Walter reaches the old house after getting lost in the 'labyrinth of narrow and dirty streets which lies in the immediate vicinity of the north-western angle of Smithfield-market'.[156] To her middle-class eyes, it seems impossible that 'human beings could dwell in such fetid and unwholesome dens', and she beholds in shock the 'labyrinth of dwellings whose very aspect appeared to speak of hideous poverty and fearful crime'.[157] The twice-repeated word 'labyrinth' strengthens the feeling of a twisted, threatening space unknown to the middle-class character, who is emotionally overwhelmed and almost killed by it. The only reason Eliza survives the fall is that she breaks conventions on many levels: she is a gender-shifter and, although good at the core, she is an unwitting party in a criminal scheme. Being far more nuanced than the typical helpless heroine of Gothic and popular fiction, she survives both the fall and the dreadful tour of the slums that ensues.

Similarly, Richard loses his way in the maze of Bethnal Green before falling into the trapdoor in the Resurrection Man's house. Ironically enough, he gets lost immediately after a heated discussion with Tidkins, in which he announces that he is no longer willing to be blackmailed. Richard declares: '[o]ur ways lie in different directions, both at the present and in the future' and that he 'shall know how to be upon [his] guard' against Tidkins' revenge.[158] Eventually, he is incapable of putting either statement into practice: he takes precisely the way that leads to the Resurrection Man's neighbourhood, where Tidkins follows him, catching him off guard. As Richard loses his way and penetrates deeper and deeper into Bethnal Green, the narrator reiterates that '[t]he district of Spitalfields and Bethnal Green was totally unknown to Markham'.[159] Of course, it is midnight, the streets are 'nearly deserted', and '[t]he lamps [...] only ma[ke] darkness visible, instead of throwing a useful light upon the intricate maze of narrow thoroughfares'.[160] While Richard meant to reach Shoreditch and there find a coach home, '[e]merging from Brick

Lane, he crossed Church Street, and struck into that labyrinth of dirty and dangerous lanes in the vicinity of Bird-cage Walk', where he soon realizes that he has taken the wrong way, as he 'flounder[s] about in a long narrow street, unpaved, and here and there almost blocked up with heaps of putrescent filth'.[161] As in Eliza's case, the narrative builds up the feeling of a dangerous and claustrophobic space, which is compared again to a 'maze' and a 'labyrinth', 'long and narrow', 'blocked up' with filth. The East-end Londoner would have recognized Markham's mistake: after coming out of Brick Lane, he should have turned left towards Shoreditch, instead of crossing Church Street and straying north into Bethnal Green. The narrative shows yet another middle-class character lost in uncharted territory, a terra incognita that is, by contrast, the natural environment of the Resurrection Man. Through Richard's eyes, the East End becomes an unfamiliar, dangerous space for the reader as well, also because of the presence of Anthony Tidkins, a criminal whose deadliness transcends social and spatial boundaries. His presence is ominously suggested by the 'footsteps behind [Richard]', which appear and disappear so that Richard concludes he is either deluding himself or, worse, 'the person whose steps he heard stopped when he did'.[162]

If Bethnal Green is a labyrinth, Anthony Tidkins is the monster that haunts it, its very own Minotaur. The appearing and disappearing footsteps suggest the chilling possibility that the he may have his own private path in the maze, perhaps some hidden trapdoor in an empty house, which allows him to stalk his prey unnoticed, a perfect fantasy of underground cellar space dominion, as Bachelard theorizes it.[163] This suggestion, which seems to find confirmation in Richard's discovery of the tunnel connecting Tidkins' house to the one next to it, alludes to a subterranean labyrinth that reflects the one above the ground, connecting the Resurrection Man's narrative space in Bethnal Green with the geospace of Bethnal Green inhabited by Bishop and Williams. As mentioned in the previous chapter, the 'Italian Boy' case originated the myth that 'a warren of cellars and subterranean passages' connected the two cottages in Nova Scotia Gardens, which, Wise contends, is a powerful example of the belief that London criminals could 'move around unseen along secret pathways of their own making'.[164] Although never mentioned explicitly, Bishop and Williams hover threateningly over the narrative when Richard, unnerved by the sound of footsteps behind him, reflects on the infamous events that established the notoriety of the neighbourhood in which he is walking. He remembers stories about 'the mysterious disappearance of persons in the east end of the metropolis', and recent 'fell deeds of crime' that have been discovered 'in the very district where he [is] now wandering'.[165]

Therefore, the narrative replica of the Bethnal Green geospace is charged with the negative memory of the London Burkers' homicides. As in *Sweeney Todd*, this connects the narrative space to anxieties about disappearing below the ground and being there dismembered, that is, dissected. The scene in the Resurrection Man's house reinforces this connection: Richard, who is playing dead in the hope of saving his life, listens to the Buffer recalling a previous victim that they sold to the same 'sawbones', i.e. surgeon, who commissioned the theft of the body of the girl in Shoreditch. When they 'caught hold of a feller that was walking so comfortable along, looking up at the moon', the Cracksman had the idea of 'holding him with his head downwards in a tub of water [...] until he was drowned'.[166] The Buffer exclaims proudly:

> and now we've reduced it to a reg'lar system. Tub of water all ready on the floor – hooks and cords to hold them chaps' feet up to the ceiling; and then, my eye! there they hangs, head downwards [...] like the carcasses in the butchers' shops [...][167]

This detailed description marks the kinship of the fictional burkers and the London burkers, and reiterates images of dissection and medicine that date back to the pre-Anatomy Act era: the 'hanging carcasses' of the burkers' victims and the reference to butchery recall such popular representations of anatomy as Heath's print *A Few Illustrations for Mr. Warburton's Bill*. Markham escapes the fate of the unknown corpse he saw in the kitchen in the attempt to discover his name and address from the Mummy and prevent its anatomy sale. The narrative therefore connects the idea of identity, and losing one's identity in the maze, with death and dissection, intermingling it with the anxieties attached to the underground space.

The works for Brunel's Tunnel first, and for the London Underground later, disrupted the foundations of the city for the best part of the century, stirring atavistic fears connected with the subterranean space as the space of death and dark forces.[168] When *Mysteries* was being issued, the Tunnel under the Thames had been open only two years, after almost twenty years of intermittent construction. When it opened, Charles Pearson suggested the possibility of creating an underground railway.[169]

His idea was derided; nevertheless, it was discussed recurrently in the subsequent decades, and its detractors adopted in their discourses a range of images that spanned from hellish cosmology, to lampooning humour, to catastrophic conjectures about tunnels collapsing by the sheer weight of the streets above them.[170] Pike notes that the middle class pictured movement below the ground as dangerous, an unlikely alternative to orthodox channels of urban movement, because they perceived the subterranean space to be unsafe and uncivilized.[171] Flanders similarly argues that representations of the slums as unmappable and unsafe originated in middle-class perceptions of that space, which rested on the physical, economic, and cultural distance between the middle and the lower classes.[172] Therefore, the representation of the slum space as terra incognita, threatening the physical and moral integrity of the middle-class individual, reflected the negative moral connotations attributed to both the subterranean space and the lower class.

These concepts emerge forcefully in the spatial representations of *Mysteries*, which is, after all, a sensational narrative written by an essentially middle-class, though Chartist, author. Eliza describes the people of Smithfield as a mass of 'ferocious-looking men and brazen-faced women', who cheer as they watch two women fighting 'like wild cats'.[173] The inhabitants of the rookery that she unwillingly ends up visiting consists of 'males and females evidently of the most wretched description', as girls are 'almost naked, without shoes, or stockings' and the men are 'hatless and shoeless', which clashes with their apparent 'boisterous mirth'.[174] Similarly, the neighbourhood of Wapping where Richard emerges after his escape from the Resurrection Man's lair 'swarm[s] with crimps ever ready to entrap the reckless and generous-hearted sailor', and is plagued with miserable pubs selling 'vile adulterated beer' to 'poor half-starved coal-heavers whose existence alternates between crushing toil and killing intoxication'.[175] Finally, Ellen Monroe's descending path to ruin also happens when misfortune confines her to Golden Lane in St Luke, yet another labyrinthine spot immediately west of Shoreditch. The East End thus also becomes the space of the moral fall, where the individual is dragged towards the deadly pit from a geographical, social, and spiritual perspective.

The quintessentially inexplicable and confusing nature of the maze pervades the space of the East End in *Mysteries*, connecting its superficial and subterranean environments via the negative values of criminality and moral and physical decay. The labyrinthine slum becomes the space of the physical, social, and moral fall where only poverty provides a map, and therefore a chance of survival, sustaining the alternative knowledge that emerges in the narrative as a reaction to the unfair social dynamics that characterized both fictional and geospatial London. The social structure affected the distribution of the population in the geographical space of the city, progressively segregating the poor in overcrowded slums that jeopardized physical health and annihilated personal dignity. Resurrectionists, first, and the Anatomy Act later, protracted annihilation after death. Consequently, *Mysteries* represents the space of the slum as one that tests both the regular dweller and the occasional (middle-class) visitor.

Nevertheless, even the unknown and unmappable labyrinth has the function of mapping out the world for the reader. *Mysteries*' vertical socio-geographical space has its geospatial correspondent in the Victorian city and reflects its clearly delimited cultural boundaries. Furthermore, the narrative explicitly connects the East End space as the space of the actual and metaphorical fall of the individual, which envisages death as the ultimate outcome, to dangers related to anatomy and dissection. In the East End labyrinth, both residents and unwitting foreigners become potential preys of the burkers, hence candidates for the scalpel. Similarly to Mark Ingestrie in *Sweeney Todd*, characters precipitate in the pit below the ground of the city, where their fate is annihilation. Yet, again similarly to Mark, all three of the middle-class characters that experience the labyrinth and the fall in *Mysteries* are allowed to survive. None of them can change the situation they witnessed: Mark cannot find his way back to Bethnal Green, Eliza can only dismally relate what she saw in Smithfield, while Ellen strives to collect the fragments of herself and rebuild her identity after moving out of Golden Lane. However, by representing the socio-geographical fall of the characters and allowing them to ascend from it and relate their experience, the narrative invites the readers to use the knowledge thus provided to their advantage. In so doing, the narrative performed the educational purpose Reynolds intended for his fiction, exposing the social, political, and geographical injustices of the management of the metropolis, which influenced the disposal of an individual's remains. Dismemberment and the effacing of one's identity were indissolubly connected to the spaces of poverty, and the specular extension above and below the ground of *Mysteries*' London symbolizes the merging of social and physical death in the Anatomy Act. Hence, *Mysteries* created for its readers a map to navigate their own apparently illogic space, tracing its origins in the power unbalance that characterized their society and inciting them to act.

Although the series never explicitly mentioned the Anatomy Act, the story discussed it and problematized its consequences on the lower strata of society, conforming to the Chartist principle of instructing the masses to bring them closer to respectability and power. To this end, *Mysteries* exploited familiar figures from the folklore of the city, rewriting them in a sensational narrative. Their anachronistic presence in the almost-contemporary fictional London of *Mysteries* highlighted the inadequacy of the system in place and its violation of the right of the poor to a dignified conclusion of their life. To reinforce this point, the narrative discussed the power roles characterizing the discourse around anatomy and dissection, systematically representing the poor as voiceless in a debate they do not possess the code to understand. The code is the language of legality, which the narrative posed as the language of the oppressor, a tool used against the poor to exclude them from the conversation. As a reaction, the poor develop their own alternative code, which grants them some power within the boundaries of the space of poverty. This linguistic gap in the narrative corresponded to the actual inability of the Victorian poor to understand the language of the Act which, as historians note, put the power of deliberating over the disposal of their own remains beyond their reach. The replication of the geospace of medicine and murder in the narrative and the obsessive preoccupation with the physical and social fall into the unmappable labyrinth were functional to the critique of the exclusion of the poor from respectability, even in death. Simultaneously, these representations catered to eminently middle-class anxieties about loss of respectability, about the fear of sharing the same fate of the powerless, dispossessed pauper.

The only exception to the segregation of spaces and discourses was the Resurrection Man, the quintessential monster of anatomy, whose anachronistic presence threatens equally his social peers and social betters. His reiteration of crimes that law should have stopped and his uncanny ability to transcend boundaries made him the embodiment of the terrible possibilities the inadequate system presented. His cadaveric presence suggested that the existence of a law that, disregarding the dignity of the weaker portion of the population, would rebound on the whole society without distinctions.

Notes

1. Reynolds's Miscellany, I: 199–200. Quoted in Ian Haywood, *The Revolution in Popular Literature* (Cambridge: Cambridge University Press, 2004), 172.
2. Ian Haywood, *The Literature of Struggle—An Anthology of Chartist Fiction* (Aldershot: Scholar Press, 1995), 4.
3. His chief contemporary opponent was certainly Dickens, whose *Household Words* was meant as a 'respectable' alternative to *Reynolds's Miscellany*. Haywood, *The Revolution in Popular Literature*, 171.
4. Virginia Berridge, 'Popular Journalism and Working-Class Attitude 1854–1886: A Study of *Reynolds's Newspaper Lloyd's Weekly Newspaper* and the *Weekly Times*,' 2 vols, Diss (University of London, 1976).
5. Raymond Williams, 'Radical and/or Respectable,' in *The Press We Deserve*, ed. Richard Boston (London: Routledge and Kegan Paul, 1970), 14–26.
6. Haywood, *The Revolution in Popular Literature*, 172–73.
7. Anne Humpherys, 'G.W.M. Reynolds: Popular Literature & Popular Politics,' *Victorian Periodicals Review* 16, no. 3/4 (1983): 83.
8. Haywood, *The Revolution in Popular Literature*, 173.
9. Ruth Richardson, *Death, Dissection and the Destitute* (Chicago and London: The University of Chicago Press, 1987), 270–71.
10. Ibid., 237.
11. For a detailed analysis of the case, see Elizabeth T. Hurren, *Dying for Victorian Medicine—English Anatomy and Its Trade in the Dead Poor, c. 1834–1929* (London: Palgrave Macmillan, 2012), 7–19.
12. Ibid., 14.
13. Ibid., 19.
14. George William Macarthur Reynolds, *The Mysteries of London, Vol. 1* (London: George Vickers, 1845), 72.
15. An early version of my analysis of the Resurrection Man's monstrosity appeared in Anna Gasperini, 'Anatomy of the Demons—The Demoniac Body Dealers of the Penny Bloods,' *Journal of Supernatural Studies* 2, no. 2 (n.d.): 133–47.
16. Ibid., 138.
17. Reynolds, *The Mysteries of London, Vol. 1*, 72.
18. Gasperini, 'Anatomy of the Demons,' 138.
19. Reynolds, *The Mysteries of London, Vol. 1*, 71.
20. Ibid.
21. Gasperini, 'Anatomy of the Demons,' 138.
22. *The Trial, Sentence and Confessions of Bishop, Williams and May*, 49.
23. Cooper, *The Life of Sir Astley Cooper, Bart, Vol. I*, 373–77.

24. Hurren, *Dying for Victorian Medicine*, 83; Chris Baldick, *In Frankenstein's Shadow—Myth Monstrosity, and Nineteenth-Century Writing* (Oxford: Clarendon Press, 1987), 142.
25. In George Alfred Walker, *The Second of a Series of Lectures... on the Actual Conditions of the Metropolitan Graveyards* (London: Longman, Brown, Green and Longmans, 1847), 15, Walker reports the witness of John Eyles, gravedigger in Portugal Street cemetery, who caught his colleagues disinterring the body of his own father. Also, Walker writes how the gravediggers of Spa Fields cemetery speculated over the bodies interred there, selling teeth, women's hair, and even burial clothes, which they would 'tear off' during the night. Ibid., 23.
26. Reynolds, *The Mysteries of London, Vol. 1*, 71.
27. Ibid. Crankey Jem is an unfortunate accomplice of the Resurrection Man, who Tidkins turns over to justice to escape prosecution.
28. 'Report from the Select Committee for Anatomy' (London, 1828), 71.
29. Ibid.
30. On the other hand, the events in the narrative unfold after the passing of the Act, and, as Tidkins himself states, he 'can't get orders for the stiff'uns as [he] used to do'. Reynolds, *The Mysteries of London, Vol. 1*, 114. Consequently, the series shows the resurrectionists turning more and more towards the other crime traditionally related to bodysnatching.
31. Ibid., 122–25.
32. Sarah Wise, *The Italian Boy—Murder and Grave-Robbing in 1830s London*, Pimlico 20 (London: Jonathan Cape, 2004), 185–87.
33. In ibid., 30, Wise writes that Bishop, the London burker, was also a resurrectionist; however, there is no record that he was a thief. His associate and son-in-law, Williams, had repeatedly served time in prison for theft, but he had only recently entered the bodysnatching profession when the 'Italian Boy' case exploded. Ibid., 47–51. Furthermore, with only one exception for the theft of glass from a carriage window, no one in the gang of Ben Crouch was a professional thief. See Richardson, *Death, Dissection and the Destitute*, 70. In general, 'the evidence for an association between bodysnatching and crime seems largely to have been based on hearsay'. Ibid.
34. Reynolds, *The Mysteries of London, Vol. 1*, 125–28.
35. See Gasperini, 'Anatomy of the Demons,' 140.
36. Reynolds, *The Mysteries of London, Vol. 1*, 126.
37. Gasperini, 'Anatomy of the Demons,' 140.
38. Reynolds, *The Mysteries of London, Vol. 1*, 328–31.
39. Ibid., 330.
40. Gasperini, 'Anatomy of the Demons,' 140.

41. Reynolds, *The Mysteries of London, Vol. 1*, 127.
42. Ibid., 126.
43. Gasperini, 'Anatomy of the Demons,' 141.
44. Camden Pelham, *The Chronicles of Crime, or: The New Newgate Calendar. Being a Series of Memoirs and Anecdotes of Notorious Characters Who Have Outraged the Law of Great Britain from the Earliest Period to the Present Time*, Vol. 2 (London: Thomas Tegg, 1841), 166, 182, https://doi.org/10.15713/ins.mmj.3.
45. Ibid., 179.
46. Ibid., 180.
47. Sara Hackenberg, 'Vampires and Resurrection Men: The Perils and Pleasures of the Embodied Past in 1840s Sensational Fiction', *Victorian Studies* 52, no. 1 (2009): 73.
48. Wise, *The Italian Boy*, 287.
49. Hackenberg, 'Vampires and Resurrection Men,' 73.
50. Ibid., 71.
51. 'Report from the Select Committee for Anatomy,' 72.
52. Sally Powell, 'Black Markets and Cadaverous Pies: The Corpse, Urban Trade and Industrial Consumption in the Penny Blood,' in *Victorian Crime, Madness and Sensation*, ed. Andrew Maunder and Grace Moore (London: Ashgate, 2004), 46.
53. Horace Montgomery, 'Resurrection Times,' *The Georgia Review* 43, no. 3 (1989): 532.
54. Gasperini, 'Anatomy of the Demons,' 142.
55. 'Report from the Select Committee for Anatomy,' 18.
56. Hackenberg, 'Vampires and Resurrection Men,' 73.
57. Gasperini, 'Anatomy of the Demons,' 143.
58. Reynolds, *The Mysteries of London, Vol. 1*, 125.
59. Ibid., 126.
60. See Lisa Rosner, *The Anatomy Murders* (Philadelphia: University of Pennsylvania Press, 2010), 36–37.
61. Reynolds, *The Mysteries of London, Vol. 1*, 126.
62. Ibid.
63. R.H.M., 'The Difficulties of Dissection,' *The Times*, 1, 1858, 9.
64. Hurren, *Dying for Victorian Medicine*, 4.
65. Reynolds, *The Mysteries of London, Vol. 1*, 323–26.
66. Ibid., 323.
67. Ibid., 323–24.
68. Ibid., 324.
69. 'Police,' *The Times*, February 26, 1845, 8.
70. Ibid. Pitch was used to insulate coffins. In George Alfred Walker, *Burial Ground Incendiarism, or, the Minute Anatomy of Grave-Digging in*

London (London: Longman, Brown, Green and Longmans, 1846), 14, Walker states that the great amount of pitch found in the chimney of the Spa Fields bone-house proved that coffins burned there regularly.

71. 'The Spa-Fields Burial Ground,' *The Times*, March 5, 1845, 7.
72. In Lee Jackson, *Dirty Old London—The Victorian Fight Against Filth* (New Haven and London: Yale University Press, 2014), 123, Jackson remarks that the Spa Fields case 'was something of a hollow victory for Walker and [the MP William] Mackinnon': it yielded no concrete results in terms of legislation, nor did it cause any exemplary punishment to befall the staff that could function as a deterrent against future transgressions. Most significantly, Jackson notes, Spa Fields itself was not closed.
73. George Alfred Walker, 'The First of a Series of Lectures... on the Actual Conditions of the Metropolitan Graveyards' (London: Longman, Brown, Green and Longmans, 1849), 17.
74. Reynolds, *The Mysteries of London, Vol. 1*, 234.
75. Ibid. This detail represents a further link with the Spa-Fields case: the *Times* issue of February 26, 1845 reports the statement of Catherine Murphy, a Clerkenwell denizen who spotted the gravediggers lifting a corpse by the hair. 'Police,' 8.
76. Reynolds, *The Mysteries of London, Vol. 1*, 234.
77. Ibid., 235.
78. Walker, *The Second of a Series of Lectures*, 26. Walker's emphases.
79. Ibid., 26.
80. Reynolds, *The Mysteries of London, Vol. 1*, 325. Reynolds's emphases.
81. Ibid., 324.
82. Ibid., 325.
83. Ibid.
84. Encyclopædia Britannica Inc, 'Ghoul,' *Encyclopaedia Britannica. Britannica Academic*, 2015. Poe's description of the ghouls appears in the poem 'The Bells' (1849).
85. Reynolds, *The Mysteries of London, Vol. 1*, 325.
86. Ibid.
87. Ibid., 326.
88. Ibid., 2.
89. Stephen James Carver, 'The Wrongs and Crimes of the Poor: The Urban Underworld of the Mysteries of London in Context,' in *G.W.M. Reynolds—Nineteenth-Century Fiction, Politics and the Press*, ed. Anne Humpherys and Louis James (Aldershot: Ashgate, 2008), 153.
90. Crone, *Violent Victorians—Popular Entertainment in Nineteenth Century London*, 190–91.
91. Haywood, *The Literature of Struggle*, 1.

92. Ibid., 4.
93. Ibid.
94. In Haywood, *The Revolution in Popular Literature*, 171, Haywood notes that the antagonism of 'commentators and critics' towards Reynolds and the 'fusion of popular politics and fiction' that characterised his work lasted from Reynolds's times until recently. Dickens, as Haywood observes, was 'Reynolds's first great enemy' in his days (Ibid.), while, as noted above, still well into the 1970s scholars such as Berridge and Williams questioned the integrity of Reynolds's purposes in writing popular fiction with political undertones. Ibid., 172–73.
95. Ibid., 173.
96. Ibid., 189.
97. Ibid., 140.
98. Ibid., 146.
99. Carver, 'The Wrongs and Crimes of the Poor,' 159.
100. Reynolds, *The Mysteries of London, Vol. 1*, 112.
101. Ibid.
102. Ibid., 112–13.
103. Ibid., 113.
104. Ibid.
105. Ibid., 114.
106. Ibid. Reynolds's emphasis.
107. Ibid., 321. Reynolds's emphasis.
108. Reynolds, *The Mysteries of London, Vol. 1*, 167.
109. Ibid., 168.
110. Ibid., 170.
111. Ibid., 172.
112. Ibid.
113. Ibid., 171–76.
114. Ibid., 175.
115. Ibid.
116. Ibid., 170.
117. Richardson, *Death, Dissection and the Destitute*, 208.
118. Reynolds, *The Mysteries of London, Vol. 1*, 313.
119. Ibid., 314.
120. Ibid.
121. Dictionary of the Scots Language, "Body' n.,' *Dictionary of the Scots Language*, 2004, http://www.dsl.ac.uk/entry/snd/body.
122. This is yet another connection between the Globe Lane episodes and the Spa-Fields case: on March 20, 1845 the *Times* reported that a straw-bonnet maker named Brown 'was in the habit of officiating as a clergyman, and perform the funeral services', upon request of the managers. 'Police,' 8.

123. Reynolds, *The Mysteries of London, Vol. 1*, 316.
124. Ibid.
125. Hurren, *Dying for Victorian Medicine*, 12.
126. Reynolds, *The Mysteries of London, Vol. 1*, 315.
127. Ibid., 324.
128. Ibid. Reynolds's emphasis.
129. Anne Humpherys, 'Generic Strands and Urban Twists: The Victorian Mysteries Novel,' *Victorian Studies* 34, no. 4 (1991): 461.
130. Ibid., 457.
131. Humpherys, 'Generic Strands and Urban Twists,' 462.
132. Haywood, *The Revolution in Popular Literature*, 176.
133. Peter Brooks, *The Melodramatic Imagination: Balzac, Henry James, Melodrama, and the Mode of Excess* (New Haven and London: Yale University Press, 1995), 40. Quoted in ibid.
134. Richardson, *Death, Dissection and the Destitute*, 239.
135. Hurren, *Dying for Victorian Medicine*, 21.
136. Reynolds, *The Mysteries of London, Vol. 1*, 324. My emphasis.
137. Anne Humpherys, 'The Geometry of the Modern City: G.W.M. Reynolds and "The Mysteries of London",' *Browning Institute Studies* 11 (1983): 72.
138. Barts Health, 'Our History,' *Barts Health NHS Trust*, 2015, http://www.bartshealth.nhs.uk/our-hospitals/the-royal-london-hospital/our-history/.
139. Hackenberg, 'Vampires and Resurrection Men,' 64.
140. Reynolds, *The Mysteries of London, Vol. 1*, 311.
141. Haywood, *The Revolution in Popular Literature*, 179.
142. George Alfred Walker, *The Grave Yards of London: Being an Exposition of the Physical and Moral Consequences...Connected with Our Unchristian and Pestilential Custom of Depositing the Dead in the Midst of the Living: With the Examinations of the Author upon This...Subject Bef* (London: Longman, 1841), 9.
143. Ibid.
144. Reynolds, *The Mysteries of London, Vol. 1*, 326.
145. Walker, *The Grave Yards of London*, 19. Walker's emphasis.
146. 'BURIAL of the DEAD in the METROPOLIS,' *The Times*, October 28, 1841, 6. Regarding the body-bugs, the earliest reference I could find in the *Times* was in a letter to the editor titled 'The Dead versus the Living' and signed 'ANTI-PESTILENCE'. The writer, who claimed to 'have heard [...] from authority [they] cannot question that a peculiar insect has been generated in this place of corruption, crawling through the boards, and spreading themselves over the chapel'. ANTI-PESTILENCE, 'The Dead Versus The Living,' *The Times*, 1841, 4. I could not find the

precise passage from *Grave Yards* in the *Times*; however, in Walker, *The Second of a Series of Lectures*, 16, Walker reports the statement of the witness Samuel Pitts, who describes 'some insects, similar to a bug in shape and appearance, only with wings, about the size of a small bug'. While we cannot be sure whether or not Reynolds ever read *Grave Yards*, it is likely that he read, or heard, about the body bugs elsewhere.

147. Reynolds, *The Mysteries of London, Vol. 1*, 326.
148. Ibid., 4.
149. Bachelard, *The Poetics of Space*, Chapter 3.
150. Ibid., 7.
151. Ibid. Reynolds's emphasis.
152. Ibid., 127.
153. Ibid., 129.
154. Ibid., 128.
155. Ibid., 129.
156. Ibid., 3.
157. Ibid., 3–4.
158. Ibid., 122.
159. Ibid.
160. Ibid.
161. Ibid.
162. Ibid.
163. Bachelard, *The Poetics of Space*, Chapter 3.
164. Wise, *The Italian Boy*, 280.
165. Reynolds, *The Mysteries of London, Vol. 1*, 123.
166. Ibid., 124.
167. Ibid.
168. See Wise, *The Italian Boy*, 287–88.
169. Pike, *Subterranean Cities—The World Beneath Paris and London 1800–1945*, 33.
170. Ibid., 33–35.
171. Ibid., 36.
172. Judith Flanders, *The Victorian City—Everyday Life in Dickens' London* (London: Atlantic Books, 2013), 183–84.
173. Reynolds, *The Mysteries of London, Vol. 1*, 23.
174. Ibid., 23–24.
175. Ibid., 129.

List of Works Cited

ANTI-PESTILENCE. 'The Dead Versus the Living.' *The Times*. 1841.
Bachelard, Gaston. *The Poetics of Space*. Kindle. New York: Penguin Books, 1964.

Baldick, Chris. *In Frankenstein's Shadow—Myth Monstrosity, and Nineteenth-Century Writing*. Oxford: Clarendon Press, 1987.

Barts Health. 'Our History.' *Barts Health NHS Trust*, 2015. http://www.bartshealth.nhs.uk/our-hospitals/the-royal-london-hospital/our-history/.

'BURIAL of the DEAD in the METROPOLIS.' *The Times*. October 28, 1841.

Carver, Stephen James. 'The Wrongs and Crimes of the Poor: The Urban Underworld of the Mysteries of London in Context.' In *G.W.M. Reynolds—Nineteenth-Century Fiction, Politics and the Press*, edited by Anne Humpherys and Louis James. Aldershot: Ashgate, 2008.

Cooper, Bransby B. *The Life of Sir Astley Cooper, Bart, Vol. I*. London: John W. Parker, West Strand, 1843.

Crone, Rosalind. *Violent Victorians—Popular Entertainment in Nineteenth Century London*. Manchester: Manchester University Press, 2012.

Dictionary of the Scots Language. "Body' n.' *Dictionary of the Scots Language*, 2004. http://www.dsl.ac.uk/entry/snd/body.

Encyclopædia Britannica Inc. 'Ghoul.' *Encyclopaedia Britannica. Britannica Academic*, 2015.

Flanders, Judith. *The Victorian City—Everyday Life in Dickens' London*. London: Atlantic Books, 2013.

Gasperini, Anna. 'Anatomy of the Demons—The Demoniac Body Dealers of the Penny Bloods.' *Journal of Supernatural Studies* 2, no. 2 (n.d.): 133–47.

Hackenberg, Sara. 'Vampires and Resurrection Men: The Perils and Pleasures of Embodied Past in 1840s Sensational Fiction.' *Victorian Studies* 52, no. 1 (2009): 63–75.

Haywood, Ian. *The Literature of Struggle—An Anthology of Chartist Fiction*. Aldershot: Scholar Press, 1995.

———. *The Revolution in Popular Literature*. Cambridge: Cambridge University Press, 2004.

Humpherys, Anne. 'G.W.M. Reynolds: Popular Literature & Popular Politics.' *Victorian Periodicals Review* 16, no. 3/4 (1983a): 79–89.

———. 'The Geometry of the Modern City: G.W.M. Reynolds and "The Mysteries of London."' *Browning Institute Studies* 11 (1983b): 69–80.

———. 'Generic Strands and Urban Twists: The Victorian Mysteries Novel.' *Victorian Studies* 34, no. 4 (1991): 455–72.

Hurren, Elizabeth T. *Dying for Victorian Medicine—English Anatomy and Its Trade in the Dead Poor, c. 1834–1929*. London: Palgrave Macmillan, 2012.

Jackson, Lee. *Dirty Old London—The Victorian Fight Against Filth*. New Haven and London: Yale University Press, 2014.

Montgomery, Horace. 'Resurrection Times.' *The Georgia Review* 43, no. 3 (1989): 531–44.

Pelham, Camden. *The Chronicles of Crime, or: The New Newgate Calendar. Being a Series of Memoirs and Anecdotes of Notorious Characters Who Have Outraged*

the Law of Great Britain from the Earliest Period to the Present Time, Vol. 2. London: Thomas Tegg, 1841. https://doi.org/10.15713/ins.mmj.3.
Pike, David L. *Subterranean Cities—The World Beneath Paris and London 1800–1945*. Ithaca and London: Cornell University Press, 2005.
'Police.' *The Times*. February 26, 1845, 8.
Powell, Sally. 'Black Markets and Cadaverous Pies: The Corpse, Urban Trade and Industrial Consumption in the Penny Blood.' In *Victorian Crime, Madness and Sensation*, edited by Andrew Maunder and Grace Moore, 45–58. London: Hashgate, 2004.
R.H.M. 'The Difficulties of Dissection.' *The Times*. March 5, 1858, 9.
'Report from the Select Committee for Anatomy.' London, 1828.
Reynolds, George William Macarthur. *The Mysteries of London, Vol. 1*. London: George Vickers, 1845.
Richardson, Ruth. *Death, Dissection and the Destitute*. Chicago and London: The University of Chicago Press, 1987.
Rosner, Lisa. *The Anatomy Murders*. Philadelphia: University of Pennsylvania Press, 2010.
'The Spa-Fields Burial Ground.' *The Times*. March 5, 1845, 7.
The Trial, Sentence and Confessions of Bishop, Williams and May. London: Steill, B., 1831.
Walker, George Alfred. *The Grave Yards of London: Being an Exposition of the Physical and Moral Consequences...Connected with Our Unchristian and Pestilential Custom of Depositing the Dead in the Midst of the Living: With the Examinations of the Author upon This...Subject Bef.* London: Longman, 1841.
———. *Burial Ground Incendiarism, or, the Minute Anatomy of Grave-Digging in London*. London: Longman, Brwon, Green and Longmans, 1846.
———. *The Second of a Series of Lectures ... on the Actual Conditions of the Metropolitan Graveyards*. London: Longman, Brown, Green and Longmans, 1847.
———. *The First of a Series of Lectures ... on the Actual Conditions of the Metropolitan Graveyards*. London: Longman, Brown, Green and Longmans, 1849.
Wise, Sarah. *The Italian Boy—Murder and Grave-Robbing in 1830s London*. Pimlico 20. London: Jonathan Cape, 2004.

CHAPTER 6

Dissection Report: Patterns of Medicine and Ethics

On 2 May 1828, anonymous bodysnatcher A.B. told the Select Committee he believed that, should bodies for dissection be sourced from workhouses and hospitals, people would finally overcome their 'prejudice' against anatomy and understand it was 'done for the good of the public'.[1] It is unclear why he should say so, as the Act would put him out of business; perhaps, he had already been assured that he would be provided for, perhaps he had earned enough to invest in another venture.[2] However, his views were exceedingly optimistic: after the passing of the Act, the pauper could do nothing but abide by its provisions, but historians have shown that the obligation to leave (and sometimes being cheated into leaving) their bodies to the anatomists did not improve their perception of dissection in the least. When I began this study, I was impressed by the recurring representation of displaced bodies and unethical doctors in penny bloods and became convinced that such a significant presence in a genre traditionally considered inconsistent, and the contemporary development of a law that problematized the relationship between the pauper's body and medicine could not be a coincidence. Nor could I find the bloods' famous love for violence and the macabre a sufficient explanation for such insistence on a topic that was so painful to its working-class readers. The analysis I developed in this book is meant to produce a more satisfactory answer—it attempts to dissect, in a way, the system of discourses about the harrowing issue of dismemberment, bodysnatching, and social power visible under the surface of the

A. Gasperini, *Nineteenth Century Popular Fiction, Medicine and Anatomy*, Palgrave Studies in Literature, Science and Medicine,
https://doi.org/10.1007/978-3-030-10916-5_6

narratives. It demonstrated, firstly, that the massive body of the penny blood genre can be analysed, with the correct tools, as a literary phenomenon inscribed within a historical, social, and artistic context much like other genres. Furthermore, it showed that even the lowly penny bloods responded to political and social phenomena that impacted on their readership, interweaving in the fabric of their narratives characters, discourses and geographies, sometimes unspeakable and incomprehensible, but always frightening, connected with medical deontology and bodily agency.

As a literary form, the penny blood undoubtedly presented a quantity of stylistic flaws: plot incoherence, authorial uncertainty, oversimplified characters, and abrupt interruptions are only the most evident. However, these flaws represent an issue only insofar one intends to examine the genre using the criteria of mainstream, i.e. middle-class, Victorian fiction: an approach that such scholars as James and Moretti have long since established to be counterproductive. The sampling, or cross-section, analysis approach I adopted in this book, conversely, uncovered continuities, resonances, and patterns in the four narratives examined, addressing complex—as in articulated, nuanced—themes such as medicine and ethics. We have seen that *Manuscripts* and *Varney* on the one side, and *Sweeney Todd* and *Mysteries* on the other, take two different stands regarding medicine. The former two represent opposite kinds of medical man, the negative monomaniac scientist and the reassuring super-doctor. These two figures act in a 'gothic' space belonging to a 'barbaric' (i.e. pre-Anatomy Act) era, where the insane doctor 'Gothicises' the space, while the heroic doctor brings logic and clarity in chaos. The latter two specimens instead do not provide a positive representation of the medical fraternity, and their urban landscapes reproduce geospaces where the underground is the 'locus' of death and dissection, but also of truth.

Besides these two distinct interpretations of the medical figure emerges the pattern that connects all four narratives, marked by ample use of popular discourses about medicine and dissection, both in terms of characters and in terms of spaces. The combination of cannibalism, butchery, mad doctors, and resurrectionists in the plots shows that the narratives engaged with the complex, broader milieu of early-to-mid-century popular culture forms, participating in popular strategies to represent and discuss lower class anxieties and beliefs about the medical world. Each narrative discusses issues relating to early-century developments in medical studies, actively questioning their ethics and highlighting the power struggle implicit in the change the Anatomy Act brought about. In so doing, they develop their own discourse about medicine, using images and tropes familiar to their audience to analyse the threat the new law posed to the integrity of their readers' bodies. Monstrous characters and images worked as a tool for displacement, as a channel to represent the fears and distress that the Anatomy Act generated in the working class, as studies in medical history such as Richardson's and Hurren's show. By doing so, all four narratives suggest to the reader a coping mechanism to face the system that threatens their right to a decent interment: the belief in the existence of positive, sympathetic medical figures (*Manuscripts* and *Varney*); the importance of discovering the truth in order to achieve closure (*Sweeney Todd*); and finally, the necessity of knowing the faults of the system in place (*Mysteries*). The sensationalized spaces of death and medicine, both real (as in replications of geospace) and metaphorical or imagined (as in the operating-theatre-like staging of Will Stephen's gaze over Clara Crofton's corpse, or Sweeney Todd's underground slaughterhouse/dissection room) provide a 'safe' arena, whose very position on the page sanctions as safe. Here, readers can face sensitive and complicated issues regarding the power that the medical fraternity (for better or worse) exercised over them. The narratives attribute slightly different meanings to these sensational spaces: in *Manuscripts* and *Sweeney Todd*, the spaces of death and medicine are the scene of a dramatic, but just, solution to the power conflicts between characters (and, vicariously, between the reader and the medical fraternity). In *Varney* and *Mysteries*, the same type of space is the stage of horror: the dramatic vampiric revivification in *Varney* and the ghastly deeds of resurrectionism and burking in *Mysteries* provide the reader with the vicarious experience of the trauma, but leave the solution of the conflict suspended.

The vertical movement I noted at the onset of my research aids the construction of this spatial discussion of the ethical and power conflict. Bodies emerge from the underground world and visit their revenge upon the characters who provoked their fall or disturbed their rest, particularly in Rymer's works. In Reynolds, the vertical movement has the twofold purpose of both commenting on the ease with which the urban space provokes the fall (physical, moral, and social) of the individual, and of discussing anxieties about bodily annihilation in the urban subterranean space. In all four narratives, class is an important element that marks the bodies that resurface from under the ground: Mary Sinclair, Long

Hannibal Jeffries, Sweeney Todd's victims (including Mark Ingestrie, the survivor), and Mrs. Smith's unknown tenant are all part of the lower orders of society, people whose right to burial without dissection was not protected by the law either before or after the Anatomy Act. Varney became the subject of Chillingworth's galvanism experiment because he was a hanged felon, and therefore belonged to a category that the popular mind traditionally associated with dissection and anatomy, even after the Anatomy Act. Clara Crofton and the unnamed, but obviously middle-class girl of 'The Body-snatchers' episode represent middle-class anxieties about suffering the fate of the lower class after death.

I would therefore state that the displaced/dismembered body resurfacing from a subterranean space in the four narratives powerfully represents the fearful possibility of becoming a subject for dissection after death, losing one's identity for the benefit of science and the wider society. The classes that manage power, that is the middle and upper classes, indeed, carelessly (and silently) sanction the sacrifice of certain bodies. The displaced/dismembered bodies and their movement through space also become a channel for a commentary on the medical fraternity, depending on how they react to the dead body and its movement, and how they position themselves in the spaces of medicine and the spaces of death (i.e. cemeteries, vaults, and rooms where corpses are laid). These images and commentaries are offered to the reader in the form of entertaining literature, of a kind that, we must remember, was located outside, and possibly against, mainstream culture. This literature circulated among working-class readers discourses from which they were excluded, as they did not belong to the class that dictated the social, spatial, and political norms that regulated the new industrial society. In this sense, the four narratives were revolutionary, not because they incited to revolution, but because they circulated topics otherwise wrapped in silence and unintelligible language among the social constituency that was interested in knowing them. By stating this, I do not mean to claim that the penny bloods purposefully, successfully, and beyond doubt instructed the working class on the Anatomy Act. I would rather assert that the simple act of *circulating* the discourse around medicine and dissection through more or less metaphorical medical figures acting alongside cadavers kept the topic alive in the minds of readers, quite possibly increasing awareness about the issue. In this way, the penny bloods here examined acted against mainstream culture in a revolutionary way, allowing their underprivileged, underrepresented, and socially powerless readership to know, think about, and perhaps even discuss the issue of anatomy and dissection, which the establishment made unintelligible for them.

In 1963, in his seminal study *Fiction for the Working Man*, Louis James argued for the necessity of approaching popular literature differently than '"classical' literature", as it required to be evaluated based on whether or not it 'succeed[ed] in its intention and function in a particular time and for a particular range of readers'.[3] Simultaneously, James continued, it is necessary to consider the position of this fiction 'within the total framework of literary achievement'.[4] While authorial uncertainty makes it very difficult to make any claims regarding intentionality in the case of penny bloods, this book shows that, as a form of popular entertainment, they participated to broader popular artistic discussions about medicine, ethics and the pauper. Some did so with overtly political intent (Reynolds), while others nonetheless showed influences from contemporary political discourses that are too blatant to be ignored (Rymer). Reaching this global view has been possible only by putting James's suggestion into practice and focusing on what the four narratives examined reveal about the 'intention and function'[5] of the penny blood genre in its time and for its readers. In this regard, the study confirms Powell's 2004 claim, reiterated to an extent by Mack in 2007, that the topic of the Anatomy Act was 'relevant' to the penny bloods' working-class audience, and expands it by making it the starting point of a larger scale analysis. The penny bloods examined, which, as noted in the Introduction, were all successful series in their time, explored the issue of dissection and medicine by deploying the market dynamics Powell explores and the connections with real events and popular culture Mack examines. However, as I have demonstrated, they also went far beyond, encompassing the wider social and political implications of the Anatomy Act. They incorporated popular tropes regarding medicine and dissection and adapted them to the strategies of the genre, producing a commentary on the ethics and dynamics of the Anatomy Act in society. In this respect, this study partially confirms and partially departs from certain conclusions Crone drew from her 2010 analysis of *Sweeney Todd* in *Violent Victorians*. My analysis demonstrates that Crone's approach to penny blood fiction as part of anti-establishment culture is crucial to understand its function towards its readers, and it extends this argument to issues of social and political power and of physical integrity, which do not apply exclusively to *Sweeney Todd*. By contrast, it suggests the impossibility of applying to the whole genre Crone's conclusion that the bloods' happy ending 'appeased' the reader but 'offered no solutions to the social problems [the readers] experienced'.[6] More crucially, Crone's conclusion that 'they did not suggest the readers should protest against the established order', but rather 'aimed at gently assimilating them into their new environment' is not universal either.[7] While *Manuscripts* and *Varney* did invite the reader to trust the reassuring image of the heroic physician, *Sweeney Todd* and *Mysteries* markedly and harshly criticized the establishment. They explicitly suggested that truth and knowledge were being controlled, hidden, and manipulated. Analysing their treatment of the medical figure and the displaced/dismembered body, the sensational narratives emerge as channels for highlighting, and criticizing, the establishment's control over the lower class individual's physical integrity. If they did not explicitly suggest that the readers should rise against the system in place, my analysis shows that neither did they invite them to simply accept their position with respect to dissection.

From these considerations emerges what is probably this book's most significant contribution, that is: showing that the penny blood genre performed, more or less declaredly, the social and political function of 'mapping out' complex issues for their readers by building an underlying discourse around social and political power. The fact that these narratives may appear unrefined to our twenty-first-century scholars' eyes should not distract us any more from this evidence. My examination of the narratives alongside a variety of popular culture items and contemporary non-fictional documents shows that they entertained a dynamic, creative relationship with their readers' everyday reality. While at first sight penny bloods seem to be merely interested in providing readers with enough violent, unsophisticated entertainment for their penny, this book shows that the writers of these texts developed their own way of discussing crucial issues in their audience's lives. Therefore, while it can be argued that this book offers but one point of view, that of medical history, it nonetheless contributes to the process of re-thinking the way scholars approach the penny blood genre.

A re-evaluation of the role of the penny blood genre necessarily entails a re-evaluation of its relationship with connected narrative forms, such as late-century gothic and the sensation novel. As noted in the introduction, Gilbert and Sparks observe that in the second half of the nineteenth century, the novel becomes the place to discuss medical issues. The monomaniacal, or otherwise ethically challenged, clinician with its distant medical gaze, but also the medical hero, observant and able to see details invisible to others, to know the unknown about their patients' bodies, appear consistently in the novel.[8] However, as Mangham notes, '[l]iterary genres do not emerge in a vacuum' and sensation fiction, which makes no exception, has been often defined by criticism a 'hybrid' of antecedent popular genres, including the penny blood.[9] Beller likewise notes sensation fiction's roots in penny blood fiction,[10] and the two genres were equally connected to the readership's health. Both Straley[11] and Gilbert[12] note that contemporary criticism ascribed to sensation fiction, which was particularly interested in nervous states and their medical exploration, detrimental effects on the readers' nerves. Detractors accused sensation novels of jolting the reader's nervous system with too much excitement. In a similar way, penny bloods, and later penny dreadfuls, were described as 'poisonous food' that affected the (young) reader's mind as much as adulterated food would affect their digestive apparatus, and as a disease.[13] The bloods (and dreadfuls) first and sensation fiction later, were believed to influence the readers' personality *and* their physical constitution. Sensation fiction and penny bloods, therefore, shared not only an interest in medical issues, but also a peculiar critical response that ascribed to the literary object the same features of the explored subject—pernicious lack of ethics for penny bloods, and provocation of neurotic states for sensation fiction.

Sensation novels developed their own form of social commentary on the medical topics that appeared in penny blood narratives. A prime example are the unethical doctors born from the pen of Wilkie Collins, father of sensation fiction. Herr Grosse from *Poor Miss Finch* (1872) and Dr. Nathan Benjulia from *Heart and Science* (1882–1883) blend discourses on inhumanity rooted in public concerns about medical practices, hubris connected to medical knowledge, lack of ethics and issues of power. Herr Grosse is a possibly well-meaning but disruptive German eye-surgeon, who operates Miss Lucilla Finch of the cataract that she developed as a toddler. The German's intervention is depicted as meddling with nature, a subversion of the natural order of things: Lucilla's blindness might well be a disability, but it is natural to her, and its elimination catalyses the sensational—and, for the young woman, harrowing—events that will ultimately lead to her falling back to blindness. As she regains her impaired condition, Lucilla regains her peace, finds the lover she had lost, and she declines the surgeon's offer to repeat the operation. The perversion of the natural order was, as my reading shows, a feature of the *Frankenstein* narrative that such popular fiction characters as *Varney*'s Dr. Chillingworth, as well as other doctors penned by Rymer, had already explored in the 1840s. *Poor Miss Finch*'s plot also pivots on the theme of sight, both of the patient and, crucially, of the surgeon: Herr Grosse, with his powerful clinical eye, can 'see' into Lucilla's blindness and assures he can restore it; however, much like Rymer's myopic Dr. Chillingworth, he is unable to foresee the consequences of his experiment and, after giving Lucilla the gift of sight, she withers, changes, and is exposed to danger. Sparks compares the perplexed Herr Grosse to 'a bewildered god' surprised by the unexpected consequences of the operation,[14] which resonates with Chillingworth's bewilderment in contemplating the revivified, terrifying Varney. Even more disquieting is *Heart and Science*'s Dr. Nathan Benjulia: enthusiastic vivisectionist, expert of diseases of the nervous system and completely devoted to the pursuit of new knowledge in the field, he is the quintessential unethical doctor. Carmina, the story's heroine, becomes isolated from her lover and, subjected to constant abuse by her unloving aunt, she develops a nervous disease. Benjulia watches her condition worsen, consciously choosing not to cure her in the hope that the disease should kill her, allowing him to perform a post-mortem on the girl's brain. While the overall discourse related to Benjulia in *Heart and Science* is anti-vivisectionist, when it is a matter of cutting bodies open, the spectre of dissection looms close by. Commenting on anti-vivisectionist discourse and Benjulia's character, Straley quotes Lewis Carroll's anti-vivisectionist pamphlet *Some Popular Fallacies about Vivisection*, which claimed that the practice of vivisection, repeated through time, must harden the man's soul against other beings' suffering and may even end up making dissection a matter of 'morbid curiosity' and even 'delight'.[15] The imagination of this disquieting distortion of the 'practice makes perfect' principle echoes early-to-mid-century public concerns about the effects of dissection on medical students, and indeed dissection is Benjulia's ultimate goal. He is the ultimately inhuman(e) practitioner, the development of figures such as Musgrove, Dr. Garratt, and *Mysteries*' doctors: like them, Benjulia disregards the dead, which to him are mere dissection material, and the ties between him and the girl's family are not an obstacle to his thirst for knowledge. He displays, however, a further degree of callousness: he disregards the patient while she is still alive, as he is unheeding of Carmina's suffering. As in Rymer's Chillingworth, the living patient is almost an inconvenience to Benjulia and his sole interest is the 'physiological' aspect of her disease.

From this perspective, both Benjulia and Grosse address another theme that penny bloods explored in their elaboration of medical ethics anxieties, that is: the power relationship between male doctor and female patient. Grosse dictates that, should the girl's family agree to his performing the operation, they shall relinquish to him complete decisional power over his patient's life during the time of convalescence. The patient must entrust completely, not simply her health, but the management of her own body, her social interactions, her movement through space, to the male doctor. Sparks notes that it is nothing but Grosse's operation that changes Lucilla from a comparatively autonomous and healthy individual into a 'patient',[16] and she also compares the touch of Grosse's lancet on Lucilla's eye and the relationship that he builds with her as a metaphor for the 'sexual conquest of the virgin'.[17] In brief, Grosse's interest for Lucilla becomes as inappropriate as that of Musgrove for Mary Sinclair, and of *Mysteries*' bodysnatching doctor for the dead middle-class girl in the Shoreditch cemetery. Benjulia too shares this inappropriateness mingled with power: alone, ill, Carmina is a powerless, female, patient in the hands of a merciless doctor who plans to penetrate her body. The inappropriateness of the medical gaze over the powerless female body, picking up on the pre-existing narrative that elaborated this theme for the penny bloods' working-class readers of the 1840s, enters the middle-class reading market. Straley argues that Grosse and Benjulia 'interrogate the cultural authority ceded to science at the end of the century'[18]; Gilbert[19] and Sparks[20] also observe that the novel in general, and sensation fiction specifically, pick up on the medical profession's increasing authority and power, which further expanded around mid-century. Gilbert, in particular, notes the rise of medical anti-heroes, or mad-doctors, near medical heroes, characterized by a 'lack of humanity' that borders in a 'god-complex'.[21] It is noteworthy that the wronged patient also changes from a pauper to a middle-class individual: the medical fraternity's increasing power meant their piercing, inquisitive clinical gaze started to focus on middle-class patients as well, hence the nervous representations of unethical and disruptive doctors in middle-class writing. The pauper, on the other hand, had long-since been the sole object of this gaze, and was aware of it, which caused their fiction to discuss the theme of the unethical/mad doctor several decades before the novel did.

Lack of ethics characterizes also doctors penned by Robert Louis Stevenson. His short story *The Body Snatcher* (1881) clearly shows that the connection between unethical doctors, bodysnatching and murder was still very much alive in the popular mind towards the end of the century. Furthermore, a proper gothic punishment befalls the resurrectionist doctors in the story: the murderer victim they had hastily dissected resurfaces whole to confront them with their deed, echoing the haunting, vindictive bodily presences of Mary Sinclair and Long Hannibal Jeffries in *Manuscripts*. Experiments and the detrimental effects of unethical medical practices are also the subject of the much more famous *Strange Case of Dr. Jekyll and Mr. Hyde* (1886). This time, the unethical medical scientist is the subject of his own experiment: exploring the double nature as powerful healers/inhuman monsters ascribed to the medical fraternity, the narrative shows a doctor who is constitutionally incapable of accepting the duality of his nature and, trying to find a solution in science, yields to his base, animalistic side. The core point is that Jekyll's quest is unethical and driven by hubris, an attempt to prove his scientific superiority and simultaneously find a way to indulge his lower instincts without fear of punishment. In yet another distorted application of the 'practice makes perfect' principle, his abuse of the chemical concoction that brings forth Hyde causes the changes to happen spontaneously and, possibly, to become permanent, leading the doctor to commit suicide. The narrative, therefore, shows the selfish pursuit of scientific achievement as ultimately dehumanizing, in the same way as penny bloods represented doctors who had lost all human sensitivity due to prolongated exposure to dissection as inhuman.

In 1897, Bram Stoker exhumed the figure of the medical hero as knowledge able but also prone to action, in the same style as *Manuscripts*' physician and *Varney*'s Dr. North, with Dr. Van Helsing in *Dracula*. In *Dracula*, the vampire assumes again its role as symbol of plague and contagion to discuss issues of blood transfusion, sexual disease, and contamination from the outside(r). Dr. Van Helsing, like Dr. North, mingles medical knowledge with a certain openness of mind that allows him to contemplate the vampire's supernatural presence. His dynamic character prevents Dracula from killing Mina, although he cannot prevent Lucy Westenra's death. Lucy, like Clara Crofton, is killed with a stake and, although her second death is a much more private event than the mangling of Clara's body at the hands of the mob, the similarity is much too evident to pass unnoticed. Other researchers have observed the similarities between certain episodes in *Dracula* and *Varney*[22]; I would venture that Van Helsing, while a much more developed character that delves deeper and more accurately than its penny-blood counterpart in contemporary medical developments, as it was custom in late-century novels, is an evolution of the active, knowledgeable, manly Dr. North. Both doctors select their (male) interlocutors to preserve peace (though one wonders, had Van Helsing explained to Mrs. Westenra why he had filled her daughter's bedroom with garlic flowers, if Lucy might not have been saved after all). Also, both are the guardians of the manliness of the men in the plot, threatened by the foreign vampire's overpowering masculinity.

This brief survey perhaps leaves out many other examples of unethical doctors in late-century novels; nonetheless, it shows the continuity existing between the characters and discourses that penny bloods first, and sensation and late-century Gothic fiction later, used to face a fast-changing medicine, increasingly powerful both scientifically socially. Medical practices became progressively invasive: anaesthesia allowed for a longer permanence of the surgeon's hand and lancet within the living, unconscious patient, and psychiatry penetrated deeper and deeper in people's mind. The medical gaze became increasingly piercing as the century progressed, and not just the poor, but the whole society started processing through literature the uncomfortable feeling this gaze awakened. However, this process started long before Wilkie Collins, in the lurid little pages of the lowly penny blood, where the taboo topic of what happened to whose bodies to achieve 'the greatest happiness for the greatest number' was addressed and, in a way, dissected. The magnitude of the penny blood corpus necessarily entails that features might appear scattered, giving an overall impression of incoherence. A cross-section analysis of representative specimens such as the one developed here, by contrast, discloses, below the exterior apparent flaws of this literary monster, a pattern that runs through the genre, made of underlying discourses and symbols that translated its readership's fears that death might prove as hard as life. In the last decade or so, the monster, which was thought to be dead, has instead spawned—and continues to do so—its terrifying offspring in current crime and supernatural-themed TV series set in the Victorian era. Showtime's *Penny Dreadful* (2014–2016) constitutes a prime example, which also explicitly refers in its title, and in several episodes, to the tradition of cheap serialized Victorian fiction. This renewed popularity constitutes an excellent occasion to open a dialogue between the public and this still comparatively unexplored literary genre, which requires scholars to adopt new perspectives and new approaches to explore its specimens. This book contributes to this process. It shows that the adventures of the heroic physician, Varney the vampire, Sweeney Todd the demon barber, and Anthony Tidkins the Resurrection Man were, among other things, the outward symptom of a broad and complex discussion about the disposal of their readers' remains. These characters spoke to the voiceless, powerless pauper who read *The Mysteries of London* and dreaded the surgeon's slab. The horrifying displaced and dismembered corpse, with the ambiguous medical man hovering over it, answered more complex needs than bloodlust and voyeurism: it was the key to a code that mapped the working-class readers' way through the cellars of their fears, up and out, through their Anatomy Act reality.

Notes

1. 'Report from the Select Committee for Anatomy' (London, 1828), 72.
2. For a detailed illustration of post-resurrectionism lives of famous London bodysnatchers, see Ruth Richardson, *Death, Dissection and the Destitute* (Chicago and London: The University of Chicago Press, 1987), 69.
3. Louis James, *Fiction for the Working Man 1830–1850—A Study of the Literature Produced for the Working Classes in Early Victorian Urban England* (London: Oxford University Press, 1963), 45.
4. Ibid.
5. Ibid.
6. Rosalind Crone, *Violent Victorians—Popular Entertainment in Nineteenth Century London* (Manchester: Manchester University Press, 2012), 190–91.
7. Ibid., 191.
8. See Tabitha Sparks, *The Doctor in the Victorian Novel*, 7–8; and Pamela K. Gilbert, 'Sensation Fiction and the Medical Context,' in *The Cambridge Companion to Sensation Fiction*, ed. Andrew Mangham (Cambridge: Cambridge University Press, 2013), 182–83, 187–88.
9. Andrew Mangham, 'Introduction,' in *The Cambridge Companion to Sensation Fiction* (Cambridge: Cambridge University Press, 2013), 1–2.
10. Anne-Marie Beller, 'Sensation Fiction in the 1850s,' in *The Cambridge Companion to Sensation Fiction* (Cambridge: Cambridge University Press, 2013), 8.
11. Jessica Straley, 'Love and Vivisection: Wilkie Collins's Experiment in Heart and Science,' *Nineteenth-Century Literature* 65, no. 3 (2010): 355–56, https://doi.org/10.1525/ncl.2010.65.3.348.
12. Gilbert, 'Sensation Fiction and the Medical Context,' 184.
13. Greenwood, 'A Short Way to Newgate,' 158.

14. Tabitha Sparks, 'Surgical Injury and Narrative Cure in Wilkie Collins's Poor Miss Finch and Heart and Science,' *Journal of Narrative Theory* 32, no. 1 (2002): 9.
15. Quoted in Straley, 'Love and Vivisection,' 355.
16. Sparks, 'Surgical Injury and Narrative Cure,' 6.
17. Ibid., 7.
18. Straley, 'Love and Vivisection,' 353.
19. Gilbert, 'Sensation Fiction and the Medical Context,' 187.
20. Sparks, *The Doctor in the Victorian Novel*, 7, 17.
21. Gilbert, 'Sensation Fiction and the Medical Context,' 187.
22. See, for instance, Rymer, *Varney the Vampyre; or: The Feast of Blood*, 703, 713.

List of Works Cited

Beller, Anne-Marie. 'Sensation Fiction in the 1850s.' In *The Cambridge Companion to Sensation Fiction*, 7–20. Cambridge: Cambridge University Press, 2013.

Crone, Rosalind. *Violent Victorians—Popular Entertainment in Nineteenth Century London*. Manchester: Manchester University Press, 2012.

Gilbert, Pamela K. 'Sensation Fiction and the Medical Context.' In *The Cambridge Companion to Sensation Fiction*, edited by Andrew Mangham, 182–95. Cambridge: Cambridge University Press, 2013.

Greenwood, James. 'A Short Way to Newgate.' In *The Wilds of London*, 158–72. London: Chatto and Windus, 1874.

James, Louis. *Fiction for the Working Man 1830–1850—A Study of the Literature Produced for the Working Classes in Early Victorian Urban England*. London: Oxford University Press, 1963.

Mangham, Andrew. 'Introduction.' In *The Cambridge Companion to Sensation Fiction*, 1–6. Cambridge: Cambridge University Press, 2013.

'Report from the Select Committee for Anatomy.' London, 1828.

Richardson, Ruth. *Death, Dissection and the Destitute*. Chicago and London: The University of Chicago Press, 1987.

Rymer, James Malcolm. *Varney the Vampyre; or: The Feast of Blood*. Edited by Curtis Herr. 2008th ed. Crestline, CA: Zittaw Press, n.d.

Sparks, T. 'Surgical Injury and Narrative Cure in Wilkie Collins's Poor Miss Finch and Heart and Science.' *Journal of Narrative Theory* 32, no. 1 (2002): 1–31.

Sparks, Tabitha. *The Doctor in the Victorian Novel*. Farnham: Ashgate, 2009.

Straley, Jessica. 'Love and Vivisection: Wilkie Collins's Experiment in Heart and Science.' *Nineteenth Century Literature* 65, no. 3 (2010): 348–73. https://doi.org/10.1525/ncl.2010.65.3.348.

Index

A. Gasperini, *Nineteenth Century Popular Fiction, Medicine and Anatomy*, Palgrave Studies in Literature, Science and Medicine, https://doi.org/10.1007/978-3-030-10916-5

N

O

P

Q

R

The manufacturer's authorised representative in the EU is Springer Nature Customer Service Centre GmbH, Europaplatz 3, 69115 Heidelberg, Germany. If you have any concerns regarding our products, please contact ProductSafety@springernature.com

Printed and bound by CPI Group (UK) Ltd, Croydon, CR0 4YY
07/10/2026
02229182-0001